Planning and Building Home Additions

Joseph Schram

Ideals Publishing Corp.
Milwaukee, Wisconsin

Table of Contents

ISBN 0-8249-6113-7

Copyright © 1982 by Joseph F. Schram

Published by Ideals Publishing Corporation
11315 Watertown Plank Road
Milwaukee, Wisconsin 53226

Editor, David Schansberg

Photo Courtesy of Georgia Pacific

Acknowledgments: The author wishes to express his deep appreciation to the three families who helped make this book possible—the Oscar Brauns, Newton Hawleys, and Sigurd Rosenlunds—who today are enjoying the results of successful home additions. Many, many hours were spent with these families as work progressed from start to finish. Special appreciation also is extended to Knight & Garrow, Inc., Galli Homes, and the many subcontractors and tradesmen who so patiently described what they were doing and why.

⌂ SUCCESSFUL
HOME IMPROVEMENT SERIES

Bathroom Planning and Remodeling
Kitchen Planning and Remodeling
Space Saving Shelves and Built-ins
Finishing Off Additional Rooms
Finding and Fixing the Older Home
Money Saving Home Repair Guide
Homeowner's Guide to Tools
Homeowner's Guide to Electrical Wiring
Homeowner's Guide to Plumbing
Homeowner's Guide to Roofing and Siding
Homeowner's Guide to Fireplaces
Home Plans for the '80s
Planning and Building Home Additions
Homeowner's Guide to Concrete and Masonry
Homeowner's Guide to Landscaping
Homeowner's Guide to Swimming Pools
Homeowner's Guide to Fastening Anything
Planning and Building Vacation Homes
Homeowner's Guide to Floors and Staircases
Home Appliance Repair Guide
Homeowner's Guide to Wood Refinishing
Children's Rooms and Play Areas
Wallcoverings: Paneling, Painting, and Papering
Money Saving Natural Energy Systems
How to Build Your Own Home

What Building Additions Is All About

Your interest in successfully adding on to your home is shared by countless thousands of homeowners each year. The dollars spent for room additions and home improvements in 1981 surpassed the amount of money spent for new home construction. More and more families are finding that expanding is the best way to fulfill family living needs, particularly with higher financing costs to consider.

In recent years many of the nation's finest homebuilding firms have either switched entirely to remodeling or increased their services to include home additions. Builder advertising in local newspapers for home additions is at an all-time high.

Projects Discussed in This Book

Three basic types of home additions have been selected as the subject matter for this book. Together they encompass nearly every phase of adding to a home. Each achieved the desires of the homeowners, who are now enjoying the more expansive living areas that were designed and constructed to their specific needs.

Numerous potential remodeling projects were appraised at the planning stage before the author selected three that would reflect everything involved, or potentially involved, in adding on. These home expansions are detailed from start to finish; they form the basis of this book, which is aimed at helping all would-be home expanders obtain needed extra living space.

For purposes of clarity, these three projects will be referred to throughout the following pages as the "Garage Conversion," the "Master Bedroom-Bathroom," and the "Major Home Addition."

Each family had different, although related, reasons for building a home addition. One family wished independent living quarters for an elderly member of the family; thus they decided on the garage conversion. Another wanted a more deluxe adult living area; they settled on a master bedroom-bathroom addition with an option for an outdoor hot-tub patio in the future. The third family wanted more total living space for two fast-growing children, resulting in a major home expansion involving family room, second-story master bedroom, and general home remodeling.

In all three instances the families were most pleased with their present address and did not wish to move to a larger home as a means of solving space problems. Each realized full well that their new addition would enhance the future sales value of their homes and felt it a good investment.

It is well to note that at the time these home additions took shape, new home sales in the immediate area, as well as nationally, were soaring in price almost as fast as interest rates were rising.

Adding on to the existing home was clearly the correct decision for each of these families. Desired day-to-day living patterns were not as disrupted as they would have been in moving to a new location; also lower interest-rate mortgages could be left intact.

The Garage Conversion The garage conversion was simply that: a transformation of a two-car garage to a large multipurpose room with bath and storage. The room is entered from the home's new private-entry courtyard and can be used as a bedroom, den-office, or general family room complete with bar, television, stereo, desk, and furniture.

The homeowner acted as his own general contractor. This $6,000 home addition resulted in the construction of a new two-stall carport with locked storage wall for garden equipment and the like. The carport was extended in one direction to meet with the end of the house, yet was kept far enough in front of the house so as not to interfere with windows. Following completion of the carport and garage conversion, the owners designed and constructed a movable-louver entry fence and gate, poured aggregate cement, and added plantings to the front patio.

Before converting the garage to a multipurpose living area and adding a two-car carport, this home was a typical ranch-style house, with most of the front yard devoted to getting the cars in and out of the garage.

The completed entry facade of the garage conversion includes a single roof line across the side of the original garage, and above the new entry gate-fence and the new two-stall carport. The totally private inner courtyard was surfaced with exposed aggregate concrete and landscaped. Slats of the fence are movable for better air control. Entry to the converted garage is by way of sliding glass doors, shown above right.

Lauan plywood paneling was used for the walls of the converted garage, and a fluorescent lighting arrangement was hung below the original garage roof beam. The floor was leveled with a new wood subfloor, which was then carpeted.

The new bath in the garage conversion formerly was a closet in the adjoining house. Compact, this room still includes a full-size tub, water closet and single-bowl lavatory. A small jalousie window below the lighting soffit provides ventilation.

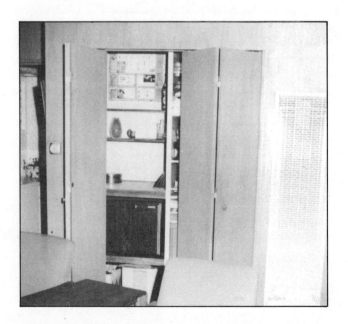

Storage cabinets were built into one wall of the garage with bifold doors used to conceal contents. The space heater installed on this wall offers ample heating comfort.

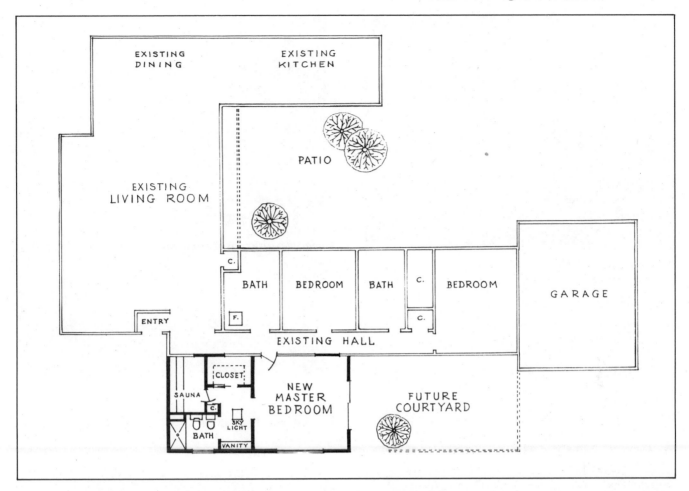

EXISTING
DINING

EXISTING
KITCHEN

PATIO

EXISTING
LIVING ROOM

C.

BATH

BEDROOM

BATH

C.

BEDROOM

GARAGE

ENTRY

F.

C.

EXISTING HALL

CLOSET

SAUNA

C.

BATH

SKY LIGHT

VANITY

NEW MASTER BEDROOM

FUTURE COURTYARD

A simplified floor plan, prepared by the contractor for the master bedroom-bathroom addition, shows the new rooms joined to the existing front wall of the home.

The Master Bedroom-Bathroom The master bedroom-bathroom is a $24,000 addition to the front of the home. It parallels a long hallway, which serves other bedrooms and baths and joins with another passageway along one side of the living room. The new addition occupies part of a former rose garden and includes a large bath, dual-bowl vanity opposite a walk-in wardrobe closet, and a family-size sauna bath built from a manufactured prefab package.

Before the addition of the new bedroom-bathroom, this home had two floor-to-ceiling double-hung windows providing light to a long bedroom-bath hallway running across the front of the house. Part of the brick and grape-stake fence was removed along with about half the rose bushes to provide space for the addition.

The master bedroom-bathroom addition is reached by a new door from the bedroom hallway and also has a sliding glass door leading to a Japanese garden that is fenced in redwood and will soon include a hot tub.

The ample-size bathroom includes a dual-person shower, bidet, and water closet. A sliding pocket door makes it possible to close this room from the adjoining dressing area.

The completed bedroom receives an interesting play of daylight; it is surrounded by a rose garden.

The dressing room area includes a double-bowl lavatory, wall-to-wall mirror, and soffit lighting by means of fluorescent tubes. Drawers and doors have finger pulls which eliminate the need for surface hardware.

The Major Home Addition The major home addition cost $40,000 and involved nearly everything that could be changed in a home approximately twenty years old. There are two new rooms: a large family room opening off the original kitchen-dining area and a second-story master bedroom suite with stall shower, dressing area, and walk-in wardrobe closet.

Major changes also were made in every existing room of the original three-bedroom, one-bath ranch style dwelling. The living room has a new tile entry door. The dining room has been revitalized with new lighting fixtures and different wall, window, and door treatments.

In the kitchen, the original appliance placement has been adjusted only slightly, but the room was entirely gutted and begun again from the original stud wall construction. As an integral part of the new family room, the kitchen now has a serving bar, new wiring, new recessed fluorescent lighting, all new cabinets and ceramic tile countertops, new appliances, flooring, and wall covering. One wall of the original kitchen was removed to open the room to the newly added family room and to give a view of the attractive circular fireplace from the kitchen and dining room.

When first constructed, this home used nonload-bearing prefabricated wood wardrobes as dividing walls between bedrooms. All of these have been removed and replaced with typical stud walls and new Sheetrock-finished closets with attractive wood-panel doors. An alcove has been created in

two of the bedrooms to serve children's desk requirements.

The third bedroom was reduced in area to provide needed space for the new stairwell to the second-story addition. The remainder of this room has been converted to den-office use with a door opening into the garage and a large storage area under the stairway. No other closet is provided.

The garage remains as the location of the home laundry. A new enclosure surrounds the hot water heater, but the washtub, automatic washer, and dryer are in their original locations. Included in the garage renovation is a double-width, electronically operated overhead door which replaces two weathered plywood bypass doors, plus new lighting and complete wall sheathing with gypsum wallboard.

The original heating system consisted of gas-fired unit heaters built into the walls in various areas of the home. All of these have been removed in favor of a gas-fired warm air system located in a closet at the top of the new stairway with ducts installed in walls throughout the first floor.

The original bathroom had been updated in recent years and just required removal of an awning window (which now would be in the middle of the family room) and addition of a new combination heating-lighting-vent fixture in the ceiling. Wall covering was replaced, and tile was extended in the tub-shower area where the window was removed.

Original doors, steel windows, and interior trim throughout the house were removed and replaced with weather-stripped wood windows, decorative

An attractive brick-and-grape-stake courtyard enclosure fronted this one-story tract home which was expanded by the owners to include a second-story master suite and rear-of-the-house family room.

The rear of the house had a plastic-panel patio roof covering a large concrete slab. The door at left opened into the dining area of the living room, while the other door at right opened into the middle of the kitchen. Window at right was above the bathtub in the home's only bathroom.

The covered entryway has a standard flush-panel door that opens immediately into the living room and another door that opens into the garage.

This kitchen window wall and part of the adjoining wall were removed in creating the new family room. All kitchen cabinets and appliances also were removed and replaced.

panel wood doors, and attractive, stained wood trim.

Prior damage to the concrete prompted a new driveway, which adjoins a new walkway along one side of the home. Future rear-yard landscaping is planned by the owners, who also improved the exterior appearance of the house with their selection of a wood shake roof replacement.

All interior asphalt tile was removed and replaced with carpeting, and all lighting fixtures were replaced with more modern units. A smoke detector system was installed at key locations on both the first and second floors.

Truly, there was not much less done to this house than would have been required to start on a fresh site and to build from the ground up. All major home building trades were involved.

Each of the three projects involved various contractors, and the two smaller improvements were partially accomplished with owner assistance. All involve typical residential construction techniques common throughout the United States, were designed with the full involvement of the owners, and were constructed to meet existing building codes and requirements.

Using This Book

Literally hundreds of hours of planning went into these home additions before contracts were signed and work begun. It was during this time that the owners fully determined specific needs and in general learned a considerable amount about the construction industry, new building materials, how the job would be handled, and many more factors that would influence their daily lives for several months of construction. But in spite of this planning, each homeowner was heard to comment many times over, "That's something I didn't think about" or "I wish the book you're planning had been written before we started, so we would know what to expect."

To aid your understanding, succeeding chapters will cover every phase of design, planning, and construction. At times the methods involved will cover what was carried out at all three homes; other times specific single installations will be detailed.

You can use these instructions as your need and experience dictate. You may wish to carry out many of the activities that do not require specialized equipment; you may wish to work out a division of services with your contractor in order to cut down on costs; or, you may wish to use the ideas and steps presented here to familiarize yourself with all the alternatives that are open to you. No matter which path you choose, you will be better able to deal knowledgeably and firmly with the people you hire for the job, and with the jobs you take on yourself.

A contractor's assistance before construction begins can be most helpful, not only in supplying answers to your many questions, but also in making suggestions that can save construction dollars and provide more livable area. Over the years many

The revised floor plan for the major home addition shows both the second-story master suite and the new family room. In the bedroom suite, the toilet and tiled stall shower are compartmentalized from the vanity-dressing room area. The home's new warm-air heating unit is located at the top of the stairs, adjacent to the shower stall.

persons who have remodeled their homes have remarked to the author about the great amount of time the contractor's salesman spent with them before the job started and the poor communication that then existed throughout the period of construction. It is recommended that you approach this problem at the outset before you sign a contract, determine just who will be responsible for what, and plan for meetings with the contractor on a regular basis during construction to iron out the many details which generally surface.

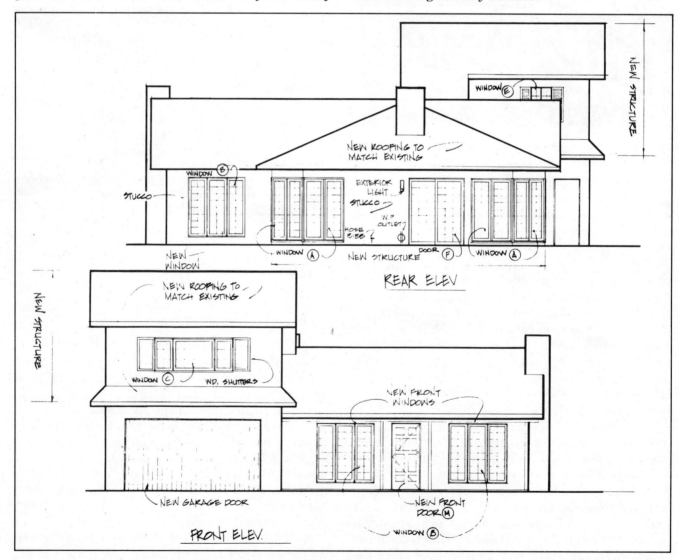

Front and rear elevations indicate the additions made to the basic tract-style home. The rear elevation shows triple-window arrangement in the existing living room and the new family room to the right. The front elevation shows the new second-story master bedroom suite and new facade for the existing lower story.

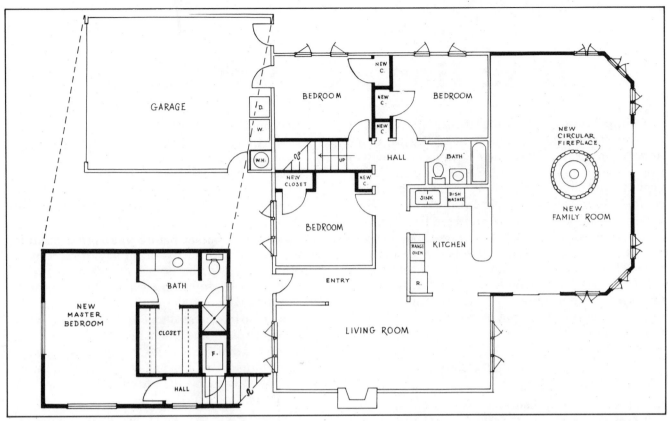

Architectural elevations prepared for the major home expansion detail the side views of the major home addition. At top is the garage side of the home, which has no windows in the new second-story master suite and the existing garage-wall window—the only window in the entire house not replaced. Below is the elevation from the living room side of the house, which indicated new additions at front and rear.

Basic Home Construction

Adding living space to an existing home results in many situations not common when building a new home from scratch. Elsewhere in this book you will find detailed instructions for specific remodeling projects, but this chapter covers some of the basics common to remodeling in general.

Begin by making an honest and realistic appraisal of your needs and just how much remodeling you can afford. Weigh these decisions in relation to the value of surrounding homes to determine if you are running a risk of overbuilding the area. If so, you will need to determine if you are willing to spend some money that may not be returned in an eventual sale of the home.

Physical Checklist

Property lines, of course, must be a factor in coming up with the final solution to your space needs. Deed restrictions, also, may impose restraints (such as a second story addition in a community of single-story houses).

Just as important in your decision-making process is the orientation of the new space, not only in relation to other existing rooms of the home, but also in relation to intended use. Many new rooms, once added, have immediately lost their true potential because of unrealistic placement for foot traffic and poor location in relation to sun and shade.

Some new home spaces should be directed toward the sun for best use of light and others away from it to prevent heat buildup. Where this is not possible, consider controlling the sun's entry to the space via sun screens, awnings, expanded overhangs, etc.

Additional Room	Suggested Orientation			
	North	South	East	West
Bedroom	3	1	1	4
Living Room	4	1	2	1
Dining Room	4	1	1	2
Kitchen	4	1	1	4
Bathroom	1	1	1	1
Garage (cold climate)	1	4	3	3
Garage (warm climate)	4	1	2	2
Laundry Room	1	4	2	2
Playroom	4	1	1	3

Suggested orientations for particular rooms: (1) excellent; (2) good; (3) fair; (4) poor.

In general, here's a rundown on the most desirable orientation of basic rooms within the home and a few pertinent reasons why:

- Bedrooms should face east or south to help conserve energy. Early-morning sun will help to heat these rooms in the winter and provide a cheery touch during the summer. By late afternoon, the rooms will have been shaded and cool enough for relaxing sleep, even in warm climates.
- Bathrooms can be placed in any direction, as the time spent here is minimal in relation to other rooms in the home. The owner's preference for early-morning sun or late-afternoon sun may well be a determining factor.
- Kitchens, like bedrooms, are best utilized when facing the south or east where they can capitalize on the sun's light-giving rays. Facing west, the kitchen often becomes a "hot house," since it is most used in the late afternoon hours for meal preparation.
- Living rooms are most usable when facing south or west and when overhangs have been planned in direct accordance with the sun, eliminating glare from sunsets. Use of large glass panels should be planned accordingly.
- Family rooms and playrooms go well on any side of the home, except in a northerly direction where the lack of sun tends to keep the room dark and cold. Here again, overhangs and extensions of the room to patios via sliding glass doors can be helpful in controlling room temperatures.
- Dining rooms are best located on the south or east side of the house, or less desirably to the west. It should be determined whether this room will be used most for family meals or for entertaining, and at what times of day the room will most likely be used.
- Utility rooms are a natural for the north side of the house because the laundry appliances create their own heat (helpful in winter and not offensive in the summer) and are used for only short periods of time.
- Garages and carports are best located on the north or south side of the home in cold or warm climates respectively. Such placements act as barriers to warm or cool the home.

With the ever-increasing need for energy conservation (regardless of the heating system to be used), local building departments are giving more attention to window placement and size. This factor de-

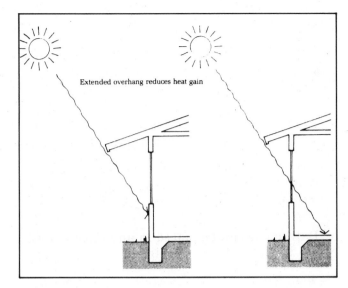

To block sun's rays from your room addition, plan an overhang that will let the sun in during winter, but not in summer. Drawing courtesy of Owens-Corning Fiberglas

serves consideration and often will be a fixed determination related to the amount of glass allowed per square foot of the space being added.

Hand-in-hand with the considerations of orientation are those of size, height, and shape. Higher ceilings require more heat and air conditioning than standard eight-foot ceilings. Likewise, long narrow home additions are harder to heat and are less desirable than a square room or rectangular space.

Keep in mind also the basic building materials you will be using. Carpeting and sheet vinyl flooring come in standard widths which, when used in their entirety, eliminate costly waste material. Remember that adding a few inches to a room dimension may incur disproportionate expense in terms of materials required.

House Components

It is well to have a basic understanding of house construction when planning an addition. This knowledge will help you to quickly eliminate the negative and accentuate the positive—saving both time and money. If at all possible, locate the original blueprints for your home, for they will convey a wealth of information as to exact location of wiring, plumbing lines, and other items now concealed in walls and floors—but which must be located for "tie in" with new lines. Your local building department may still have these plans on file.

Framing Successful room additions will normally require a joining of existing structural framework and framing for the new room. Most homes have been constructed with wood framing consisting of 2 x 4s placed sixteen inches on center for

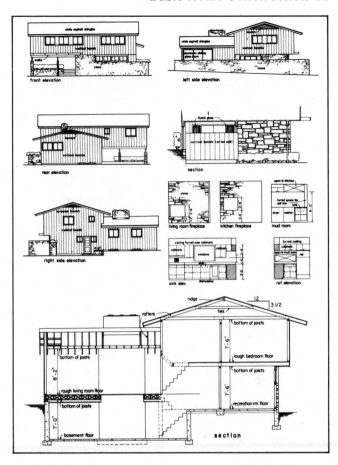

Working drawings for a home or addition to a home are called blueprints, although today these drawings are printed black on white paper. These working drawings usually contain scale drawings of each floor level, elevations of each of the four sides of the exterior, elevations of the walls of kitchen and bathroom if those are being added or remodeled, and a full cross section through the home to show portions of the structural framing. Many diagrams are color coded and can give supplementary information about operations the owner may want to handle, if requested.

walls, wood joists for floors and ceilings, and wood roof assemblies. In many areas, the home may have been built upon a concrete slab, if without a basement. Each of these factors is relatively easy to ascertain, and it is quite common to combine systems—for example, a wood flooring system with a concrete slab—when adding on to the home.

Utilities With every new addition comes the demand for utilities to service wiring, heating or cooling, and plumbing needs. Subsequent chapters of this book will detail how and why specific extensions were made to provide illumination, climate control, and plumbing. But the important initial consideration is the determination of the maximum capabilities of the existing electrical service and heating systems. Should this not afford the opportunity of "tie in," a complete new or auxiliary system may be required.

Financing and Building Permits

Arranging Financing

Two of the three home add-on projects in this book involved outside financing; the garage conversion was paid for in cash from savings over the six-month construction period. One of the finance programs required that the owner take out a second mortgage, while the third project was conventionally financed with a bank loan.

Those owners using borrowed funds for the remodeling projects shopped at several sources for the best possible loans at the most favorable interest rates, a practice recommended to anyone contemplating a home addition. Among the lending institutions checked were banks, insurance companies, savings-and-loan associations, and title and trust companies. Neither of the existing mortgages offered the possibility of "open end" financing, which would have provided funds merely by increasing the size of the existing mortgage.

In seeking a loan for home additions, you may wish to choose one of the most common types of financing:

1. A commercial loan from a commercial bank, usually repayable 3 months to a year after receiving funds. The amount of money available this way depends largely upon the borrower's income and his credit standing.

2. Commercial banks offer standard home-improvement loans, with the maximum amount loaned up to 75 percent of the home equity. The term is usually 5 to 15 years.

3. Home-improvement loans made by savings-and-loan associations have a specific ceiling; terms may run from 10 to 15 years, again with a limit of 75 percent of home equity.

4. Credit unions issuing home-improvement loans have varying policies related to the amount that may be borrowed. Most are granted for a 10- to 15-year term.

5. Institutions offering Title I, HUD-insured home-improvement loans usually will require just your signature for loans under $7,500. This financing must be obtained before work is begun; completed work cannot be financed this way. In arranging for financing, be certain of all the terms, including any early-payment penalty involved, and check to see that the interest rate quoted conforms with federal truth-in-lending statutes.

Many general contractors and remodeling firms are in a position to help their clients obtain financing. These firms may have existing relationships with various types of lending organizations that may help you obtain a better arrangement. Check with a bank, however, before signing such a loan. Be sure the terms compare favorably with those offered by other types of lenders.

Once you do sign, be certain that the loan contract provides you with the truth-in-lending three-day cooling-off period, during which time you can cancel the agreement by notifying the appropriate person in writing.

BUILDING INSPECTION RECORD
CITY OF LOS ALTOS CALIFORNIA
This is an Official Record . .To be placed in a prominent location and not to be removed except by authorized personnel.
948-1491   FRED B. CULLUM Building Inspector

Job Address ....

Building Permit No. .... Date ....

Owner ....

Builder ....

Inspector Must Sign All Spaces Pertaining to Job

TYPE OF INSPECTION	DATE	INSPECTOR	REMARKS
Location, Excavation, Forms, Foundation Steel & Materials			
POUR NO CONCRETE UNTIL SIGNED ABOVE			
Joists and/or Girders			
Rough Plumbing (filled with water)			
LAY NO SUBFLOOR UNTIL SIGNED ABOVE			
Frame (roof on and exterior enclosed)			
Rough Wiring			
Rough Gas Piping			
Flues			
Plumbing Vents Shower Pan			
Fireplace			
Stucco Wire			
DO NOT COVER INSIDE WALLS UNTIL SIGNED ABOVE			
Sewer			
Final Gas Inspection			
Final Electric Inspection			
FINAL INSPECTION (Required on all jobs.)			

Building inspection record cards issued along with building permits must be posted in a prominent location throughout home addition construction. The local building inspector signs off various work as it progresses, thereby granting approval for workmen to move on to the next phase of the project.

The homeowners constructing the major home addition first approached their savings-and-loan company that held the mortgage on the existing structure. This 7½ percent, thirty-year FHA loan had been in effect for several years and had no early-payment penalty.

The savings-and-loan officials, confronted with a tight money situation at the time the home expansion was being planned, recommended the owners take a bank construction loan with the idea of refinancing the entire package upon completion of the addition.

The owners considered this and also contacted several other banks and savings-and-loans which were charging a much higher interest rate at the time, plus five or six points. These loans also had an early-payment penalty clause.

It was the remodeling contractor who suggested the owners obtain a second mortgage from a bank that was becoming active in the remodeling market. A loan was obtained for the entire remodeling cost with a minimum set-up fee, and early-payment penalty. The term was for fifteen years at the current interest rate.

The owners now have two monthly house payments, with roughly ⅓ of the total indebtedness at the original interest rate and ⅔ at the higher rate. And there is always the possibility of combining the two notes by refinancing.

Hiring a Contractor

Perhaps the biggest factor of all in building a successful home addition is the selection of the prime contractor. It is he who will be responsible for the total job, the work of the subcontracting trades (electricians, plumbers, drywall applicators, etc.), and the payment of materials bills, as well as following specifications, dealing with city codes and inspectors, and keeping you happy with the finished addition.

Local Better Business Bureaus can show thick folders of complaints registered against the fly-by-night contractor, so this may well be an excellent place for you to check the names of those contractors you are considering. Your local building department is another checking point. And do not forget to ask for a list of the contractor's previous customers, whom you can contact both for a personal endorsement and for a first-hand look at the quality of workmanship you can expect.

Well-established and reputable contractors will be pleased to have you check them out thoroughly in advance of doing business. Likewise, these firms usually bring with them well-qualified subcontractors who can do their work correctly the first time and eliminate costly call-backs.

In signing a remodeling contract, it is well to have the instrument reviewed by your attorney. It should include total cost, specifications related to quality of materials, all of the agreements you have discussed, any guarantees you expect, and a completion date. Any and all changes made subsequent to contract signing should be committed to writing and signed by both parties. It may also be to your advantage to set a penalty clause stating what you expect if the contractor does not meet the agreed-upon completion date.

With construction costs what they are today, it is also advisable to cover your home addition with a protective bond that guarantees the contractor will faithfully perform the contract and pay all labor and material costs. In the event that the contractor defaults, the surety company has a direct obligation to take over and complete the contract or to pay the owner for his loss in cash.

The owner's protective bond usually is written for the life of the project and for the full dollar amount of the contract. This protection is invaluable should the contractor agree to do a job for a certain amount and then finds that his estimate was incorrect and the true cost will be higher. The contractor suffers the loss; should he go out of business, the surety company will make up this difference upon contractor default and then try for recovery from the contractor.

In hiring a contractor it is well to have a full set of plans and specifications for the job you wish done. These plans and specs should clearly detail every aspect of the job and leave as few questions unanswered as possible.

For example, room sizes should be accurate and all materials should be specified as to grade, model number, or specific brand whenever possible. If painting is involved, detail the number of coats you expect to receive. Anything specified in the detailed job specifications must be carried out by the contractor or professional tradesman you hire.

Money spent with an architect or designer often can result in reduced construction costs, as well as create an exact understanding between the homeowner and the contractors who will perform the services required. Try to select an architect who specializes in residential construction or home remodeling, if possible.

Doing Some of the Work Yourself If you decide to take on some of the projects—after having read through this book and acquainted yourself with the methods and effort involved—be sure to include the division of work in your contract with the contractor or subcontractors.

Prepare a letter of agreement about the work division along with the contract. Many contractors will agree to perform key or critical parts of the work and guide you in doing the balance. Or, if you are knowledgeable and experienced in construction methods, you may not even need this amount of aid. Many homeowners serve as their own contractors.

Before Construction Begins

Once you have selected the contractor or subcontractors who will undertake your remodeling project, it is well to prepare for the inconvenience you will experience during the construction. Depending upon the location of the home addition and its scope, this inconvenience can be minimal or quite extensive.

Two of the home remodelings detailed in this book brought about a minimum of household confusion—the garage conversion was begun following construction of the new carport, and the bedroom-bathroom addition did not involve the balance of the home until near the end of construction, when its doorway was cut into the bedroom-wing hallway.

More typical of the problems one can expect was the major home addition, which was begun with the family of four trying to carry on a normal existence while contractors worked around them. This involved coping with construction noises ranging from concrete jackhammers to power saws to daylong hammer pounding.

The owners in this instance resided in their home during the initial stages of construction: preparation of the foundation for the rear family room and the second-story bedroom suite addition. Then, as planned, they moved their children and belongings out of the house and gave the contractors full sway for the balance of the construction period. This factor was a financial consideration in the homeowners' favor; it gave more freedom to the workmen, who then did not have to work around furniture, children, and the like.

Regardless of whether the owner is, or is not, occupying the home during construction, certain facilities are required by the workmen, including the availability of telephone, electrical power, running water, and a lavatory or bathroom. If you will not be home during the construction day, a key must be provided so workmen can enter the house when necessary.

Permits In most instances, the contractor is responsible for obtaining all necessary building permits from local city agencies. He will need your plans and specifications to prove that the new construction meets all local requirements, but if they do not, you are the one who eventually will be held responsible unless your contract states otherwise.

In most cities, either the homeowner or contractor may obtain the building permit from the city building department. The fee varies from city to city and in many instances relates to the estimated cost of the home expansion.

In obtaining a building permit (should you not have an estimated cost for the new construction) most cities will use a rule-of-thumb amount and multiply it by the number of square feet in the addition.

Aside from the basic building permit, most communities require separate permits for electrical, mechanical, plumbing, and gas alterations and additions. Some communities base these fees on a per-item cost, such as $2 for the first 10 new outlets and switches to be added by the electrical contractor, plus a basic fee of $2 (or other amount).

Unless you are acting as your own contractor, it is advisable to have the general contractor and his subcontractors obtain the necessary permits, as they will be more qualified to answer any questions which might arise. You may have to sign the permit application in this situation.

Preparation The area to be used for expansion should be clear of plants, flowers, and other items you wish to save and use elsewhere. Do not wait until the workmen arrive to begin this chore, as it will hamper their activities and possibly result in wasted hours. Clear the site beforehand, or clean out the garage, or have the room being expanded free of furniture, carpeting, drapes, and other items.

Most contractors have become experts at getting along with children and neighborhood construction "superintendents," but the less interference by both, the better for you and the workmen. By all means, see that small children are kept away from power tools, materials, and other potentially dangerous items.

Preparing the Specifications

Ideas for home additions come from many sources, and many can be incorporated into your specific home plan. However, it is imperative that what you are thinking about is not something vague in your mind—an idea that you cannot relay to the architect, designer, or contractor who will create the necessary construction plans.

The garage conversion did not come about overnight; the owners thought about it for five years, during which time they drove through many neighborhoods seeking ideas that would enhance their home and make the eventual addition complement the original home design. Next, a detailed scale model of the entire home, converted area, and new carport was made by the owner.

The one-quarter-inch model made it possible to fully visualize what the finished conversion would do to the existing land usage, traffic patterns, home elevation, roof line, etc. The owner was determined to achieve a nonremodeled look as well as maximum use of both interior space and site.

Serving as his own general contractor, the owner prepared plans and specifications that covered what he wanted and then subcontracted heavy construction work, including beam placement in the carport, concrete work, and preparation of a new garage floor and plumbing. The carport was placed to create a new entrance court but not to interfere with existing window light and ventilation across the front of the home or along the garage-door wall of the conversion.

Plans for the master bedroom-bathroom addition were relatively simple, involving only basic blueprints and a few sheets of specifications to detail the major products involved: Sheetrock walls, plywood floor, aluminum windows and sliding glass door, prebuilt sauna, plumbing fixtures, vanity built-in, shower stall, stucco siding, and shake roof.

The major building-remodeling and home addition featured in this book were prepared with the expert assistance of an architect, put out to general contractor bid, covered by performance bond, and detailed by extensive blueprints and specifications which dictated the performance of all trades.

Foundation: Crawl Space and Slab

Foundation Preparation

Construction for a home addition actually begins with destruction: the second-story addition requires removal of part of the roof; the new room joined to the existing first floor structure probably requires removal of a wall or exterior siding; holes have to be made into walls and ceilings to extend utilities.

The major improvement program we are detailing began simultaneously at the front and rear of the home. In front, the garage roof was removed and new joists installed for the second-story addition. In the rear, a fiberglass patio roof was removed and a jackhammer was used to break up the concrete patio pad.

This removal of existing materials immediately presents a need for waste disposal, usually met by placement of a large metal container or hopper either at curbside or on the owner's property. Be sure to determine in advance if the city regulations will permit placement in the street, or if you must place the bin on your driveway or elsewhere.

Local refuse companies generally use two sizes of bins for construction jobs—one for rock and concrete, and a larger unit for wood, plaster, stucco, and other job-site debris.

Both the garage conversion and the bedroom-bathroom addition required construction of a crawl-space foundation for support of the new finished floor at the same height as existing floors throughout the house. Using a power jackhammer, a laborer first dug a U-shaped trench to accommodate dimensions of the new addition and removed excess dirt from the site. Stucco also was removed from the exterior wall that was being converted to an interior wall.

Concrete forms were then constructed of 2 x 12s, 2 x 4s, and 1 x 2 braces to hold the batter boards in place during the concrete pour. Duplex nails were used to simplify removal when stripping the forms from the finished concrete.

With the forms in place, reinforcing bars were placed approximately 3 inches from the bottom of the foundation trench and 3 inches from the top, both levels being suspended in place with bailing wire, and both levels being joined to the single existing foundation wall of the house a minimum of 3 inches into the older construction. Some local codes require three rows of bars.

Inside the perimeter foundation a dozen pier holes were dug to an 8-inch depth, 18 x 18 inches in size. These were located on 4-foot centers running parallel with the house and 6-foot centers in the opposite direction, and required no additional forming.

The shape of the perimeter foundation tapered from 12-inch-width at the base to 7 inches at the top. The total depth was 21 inches, although code required only 12 inches into the earth and 6 inches above.

Wood planks were used at key locations to facilitate the pour of concrete with a wheelbarrow. Redi-mix cement was purchased from a local batch plant, trucked to curbside and then poured in place by wheelbarrow.

In preparing foundations of this type, take special care to obtain an accurate and smooth exterior concrete surface as well as a level surface. The inside wall of the foundation need not be perfectly smooth, since finished construction will keep it from view.

As the cement is placed in the forms, a small board is used to vibrate the material, eliminating possible voids and helping the cement to settle properly within the forms. With the desired level achieved, visible wire ties used for supporting the

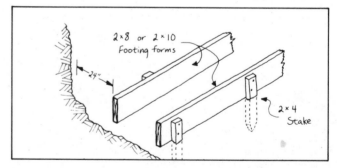

Footing forms are set parallel with top edges level and at uniform depth, measured down from the line cord stretched between batter board pairs. The actual depth will depend upon the height of foundation walls and elevation of the first floor level above the reference bench mark. Where footings of different levels meet, such as a basement area adjoining a crawl space or unexcavated area, the common practice is to end the footing at each level with ends as nearly vertical over one another as the soil condition will permit. At end points, short planks should be nailed to the parallel side forms. If form planks are set so their bottom edges are ½ inch or more above grade, avoid loss of concrete by placing earth fill on outer sides of forms to keep the concrete from flowing out.

reinforcing bars are clipped and permitted to settle into the cement.

While the cement is still fresh, 10-inch foundation bolts are pushed into the material at spacings determined by codes and construction plans. In this instance, bolts were spaced 4 feet 10 inches apart, but could have been as far apart as 6 feet, as required by code.

Each of the dozen pier holes was filled with cement, which was leveled with the surrounding ground. When cured, the concrete pads formed the base for adjustable steel floor jacks.

When the perimeter concrete footings hardened, wood forms were stripped away and 2 x 4 mudsills attached to the foundation bolts. This material first was treated with Copper Green preservative, as prescribed by code, to prevent decay.

With the mudsill in place, a 2 x 6 ledger was nailed at floor height to the original exterior wall, running the length of the new addition. Four 4 x 6 wood girders were then set parallel to the ledger, three of the rows atop adjustable floor jacks and the fourth atop the exterior mudsill. A string line assured uniform height of the members, which were then braced by the application of 2 x 4s spaced 40 inches apart. Many codes do not require this bracing, but it provides a sounder flooring system.

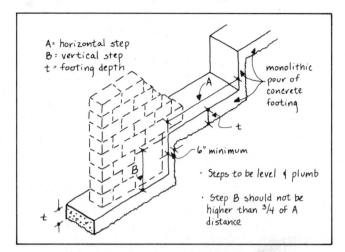

When specifying or arranging foundations, check with your local building code. Stepped footings are required to be continuous under some codes. In such cases, the earth base must be dug to grade on the different levels, with vertical sections spaded smoothly. Form boards or plywood sheets are nailed to long stakes to form the vertical wall face for the footing, and the form must be well braced to withstand the concrete pressure during pouring. If pouring stepped footings, begin at the lowest level and pour continuously. Don't let the redi-mix chute deliver concrete directly into vertical portions. Pour the concrete onto the flat area, and then shovel it down into the vertical portion.

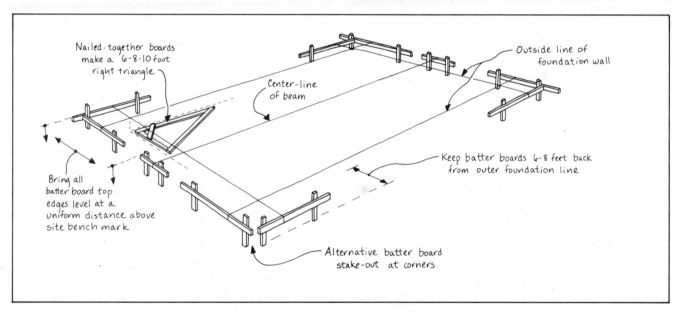

Batter boards, nailed to carefully placed stakes at building corners and midpoints, act as fastenings when stretching mason's line cords between board pairs for proper alignment and height for the rooms' excavation, footings, and foundation walls. Although the boards themselves need not be perfectly aligned, their top edges must be exactly level. This means nailing up the boards by means of a level-transit so that each board's top edge is a uniform height above the homesite's reference bench mark.

Batter boards should be set at least 6 to 8 feet back from the outer foundation wall face. Once boards have been placed and made level, line cords should be stretched and adjusted for position directly above the outer face of the foundation wall.

Then drive a nail partially into each board's top edge where the line cord crosses. Drive a second nail to the outside of the first nail by the footing projection distance. The second nails will give alignment for the footing forms. Line cords can later be removed for excavating work and easily replaced in proper position alongside the nails when setting footing forms and beginning the foundation wall.

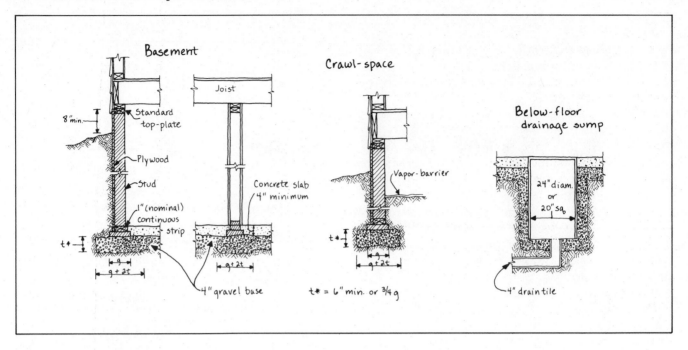

Basement

Crawl-space

Below-floor drainage sump

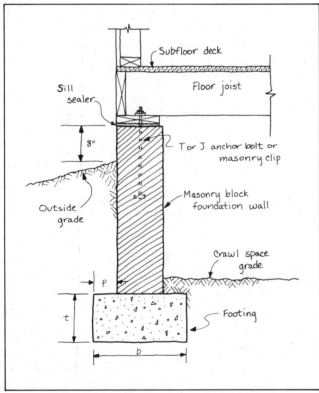

Shown is an all-wood foundation, now approved by some major building codes and by government loan-insuring agencies. They have been developed because of their high resistance, due to pressure treatments. In this sketch, the wood parts, shown by diagonal lines, are pretreated. Time and trouble can be saved by using prefabricated wall sections in panelized form.

The parts and materials involved in house foundation support are shown in this sketch. Floor construction must be anchored to foundation wall by bolts or appropriate metal clips. For concrete footings, thickness (t), base width (b), and projection from the face of the foundation wall (p), will be determined by the height of the building and whether or not a basement is planned. In the FHA Manual of Acceptable Practices, footing thickness may be 6 inches minimum for one- and two-story homes, but 8 inches on a two-story with basement if built of masonry or masonry veneer. A footing projection of 3 inches is suitable to no-basement frame construction, while a 4-inch projection is necessary for basement or masonry homes. Requirements vary according to local building codes.

A local building department inspection is required before the framing is covered with plywood or lumber. This is also the best time to install additional plumbing lines and heating ductwork, as the workmen can complete these jobs standing up rather than lying on their backs in a dark crawl space.

Various subfloor systems can be used atop girder construction, and in this situation, a 1⅛-inch tongue-and-groove plywood decking was selected.

The 4 x 8-foot panels were nailed to the girders and to intermediate members with 10d common nails. End joints of the plywood panels were staggered, leaving 1/16-inch spacing between edge and end joints and 3/32-inch for tongue-and-groove edges.

The adjustable steel floor jacks were nailed to the concrete pier pads with four No. 10 1¼-inch concrete nails. Flanges supporting the girders were nailed to these members, still permitting any necessary adjustment of the height at a later date.

1. A jackhammer was used to loosen compacted earth, which was removed from the site of the bedroom-bathroom to lower the grade to create a new crawl space area. This tool also was used to cut through the concrete foundation wall for a crawl-through opening between the old and new sections of the house.

2. While the footings were being dug, stucco was removed from the front wall of the home so that debris could be removed at the same time as the excess dirt. An axe was used to break the stucco into sections 2 to 3 feet square.

3. Aviation snips were used to cut the old wire mesh embedded in the stucco, with cuts concentrated on the vertical plane. A horizontal break then was made with the axe, without horizontal cutting with the snips.

4. A pick proved handy in pulling the broken sections of stucco from the wall, leaving only the original building paper to be stripped from the board sheathing that would become the inner wall of the new addition.

5. With the wall area stripped of stucco and the grade lowered to proper height, work could begin on preparing the concrete forming. Papers tacked to the wall included plans and city building department inspection card.

6. String guides were used to assure accurate placement of 2 x 12s used for forming the exterior side of the new footing. Material of the same dimension was placed flat on the ground and secured with wood stakes to prevent any possible movement of the exterior formwork during the concrete pour.

7. With the outside perimeter form boards completely in place, two rows of reinforcing bars were secured to these boards and the inside form boards secured in place. Note that inside formwork slopes slightly. Also, the cross brace and inside stake were removed once concrete was in place, before hardening.

9. With all the forms in place, a jackhammer was used to cut into the existing foundation wall to provide the crawl opening between the new and old sections. The plumbing and heating later were extended through this opening to the new addition.

8. Reinforcing bars used in the new footings were tied into the original foundation to a depth of 3 inches. This foundation also was chipped away on one side to provide further bonding of the new and old.

10. Redi-mix cement for the foundation footings was delivered to curbside and transferred to a wheelbarrow for movement to the formwork. One man operated the truck controls; another pushed the wheelbarrow.

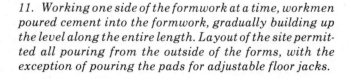

11. Working one side of the formwork at a time, workmen poured cement into the formwork, gradually building up the level along the entire length. Layout of the site permitted all pouring from the outside of the forms, with the exception of pouring the pads for adjustable floor jacks.

12. Shovels and a large stick were used to "vibrate" the cement thoroughly within the formwork, eliminating possible voids.

13. Hitting the sides of the form with a hammer further helped to settle the cement into all areas of the formwork.

14. Pads poured for support of the adjustable jacks were smoothed with a shovel; this eliminated the need for manufactured pier blocks sometimes used in place of adjustable jacks.

15. A treated mudsill equipped with foundation bolts was set atop the cement before it hardened. The bolts were pushed into the cement and maintained at the height indicated by a string line.

16. Placement of the mudsill required removal of all cross-framing members of the foundation forms.

17. Once the concrete hardened, workmen removed the stakes holding form boards and "stripped" the boards from the foundation. The 1 x 2-inch braces were secured with duplex nails to simplify removal.

A simple 2 x 4 frame was held in place with temporary stakes to guide digging of the footings. Because of the buildup of materials specified for the concrete slab, inner area earth also was removed to the necessary 12-inch depth.

Pea gravel was used to form the base of the new slab and then was covered with a 6-mil polyethylene film vapor barrier. A 2-inch layer of sand covered the vapor barrier. Note the absence of interior form boards required for crawl space footings. In slab construction, the slab and footing are poured together as an integral unit. Reinforcing bars were positioned at three levels in the footings, and the slab area was reinforced with 6 x 6-inch-No. 10 x 10 welded wire mesh, lapped a minimum of 6 inches at splices.

Special formwork was fabricated for the base of the family room's circular fireplace. A grid work of 1/2-inch reinforcing bar was set temporarily atop brick squares and raised several inches during the concrete pour.

Concrete Slab Construction

The family room addition was constructed on a concrete slab foundation which again called for a U-shaped concrete perimeter foundation joined to the existing rear-wall foundation.

Workmen lowered the grade, removing soil and preparing a pad surface of 4 inches of pea gravel, topped by 2 inches of sand, and covered with an 8-mil polyethylene film and more sand. Welded wire fabric reinforcement was then laid atop the pad and a reinforcing bar placed in the perimeter foundation excavation. Only the outside formwork was required with this system, as compared with the two-form walls required to create a crawl space system.

This slab also required special circular forming near the center of the pad to support a fireplace. Four reinforcing bars were placed in this area, rather than mesh.

Footings surrounding the new concrete pad were dug 18 inches deep, reinforced with ½-inch diameter reinforcing bar. Because of the rear yard location and the impossibility of getting a redi-mix cement truck to the placement location, a local concrete pumping service was employed.

A six-sack concrete mix was delivered to the front of the home in the customary redi-mix truck and then was poured into a small line concrete pump powered by an air-cooled 3-HP motor. A hose stretched from this unit to the rear of the home where one workman directed the flow of concrete

One workman controls the placement of concrete at the hose nozzle while a second workman helps move the hose, working approximately 8 to 10 feet from the nozzle. Footings were filled first to the height of the pad area, and then the pad was poured.

first to the fireplace formwork and then to footings and slab areas.

Use of this equipment has become common in areas where it is either difficult or costly to move the material by wheelbarrow. In this instance, better than 15 yards of cement was pumped in a fraction of the time that would have been required for wheelbarrow placement.

A 3-inch inside diameter hose conveyed the 3200 psi mixture at a rate of 1 yard every 3 minutes. This same system can be used to pump material as far as 600 feet from truck to site, but the best use is in the 250 to 300 feet range.

A pair of long 2 x 4s was used as part of the concrete formwork to simplify obtaining a level pour. Installed from the house to the fireplace pad, and from the fireplace pad to the exterior formboard, the 2 x 4s provided workmen with a guide point as they used other 2 x 4s to rough-level the pour.

After the rough leveling, workmen used long-handled wood and metal floats for finish smoothness. Treated 2 x 4s were installed at two locations in the slab to accommodate exterior doors. A narrow chase was created from the fireplace to the old exterior wall for the planned run of a gas line that would serve the fireplace starter.

Workmen next inserted foundation bolts to be used later in attaching the treated wood plate and removed excess cement from the outside of forms to simplify formwork removal.

A narrow chase was left in the surface of the new concrete slab for the purpose of extending a gas line to the new family room fireplace. This area later was covered with a wood panel over which matting and carpeting were laid.

A 2 x 4 guide secured to the top of the fireplace formwork and the exterior footing framework serves as a guide to leveling the concrete slab with another 2 x 4 screed. Workman can be seen in the background completing the actual pour of material.

Working across the entire surface of the pad, the hose operator pours a partial layer of concrete which enables another workman to raise the wire reinforcing off the sand and vapor barrier and into the forming layer of concrete.

Hand trowels and long-handled wood floats were used to finish smooth the slab. Treated 2 x 4s also were embedded in concrete at the locations of French doors.

Floor Framing

Typical floor framing in remodeling, as well as new construction, consists of sills, girders, joists, and subflooring, with all members tied together to support the expected loads and to provide lateral support to the exterior walls.

In the garage conversion, the existing concrete floor measuring 19 x 25 feet was covered with building paper to accept a system of 4 x 6s that raised the floor height to that of the rest of the house. Wood shingles were used to shim and level the support members to compensate for the slight slope of the original floor.

The 4 x 6s were then surfaced with ¾-inch tongue-and-groove plyscore which later was covered with matting and shag carpeting. In the bath area, ceramic tile was put over the subflooring, and match-ing tile on the walls and vanity.

The entire platform for the bedroom-bathroom addition was assembled with 1⅛-inch tongue-and-groove plydeck manufactured with an exterior grade glue line. These panels were installed with staggered end joints and securely nailed to the lower framework.

Floor framing for the second-story bedroom and bath above the attached two-car garage involved the removal of 2 x 10 roof joists on 24-inch centers which were replaced by 2 x 12 joists. The new members were installed parallel to the front elevation and required notching of stucco at the wall lines to permit the joist to extend as an overhang. Workmen used 2 x 12 blocking between the joists for lateral stiffness. A ½-inch plywood deck then was secured to the joists, which were spaced 16 inches on center.

A 2 x 6 ledger treated with Copper Green (for the prevention of termites, fungus rot, decay) was nailed to the front house wall. A short length of the same material was placed above the adjoining mudsill as a guide for the string level.

The 4 x 6 girders were notched to fit atop exposed foundation bolts securing the mudsills. Again a string level determined accurate height.

Blocking of 2 x 6 material was toe-nailed into the girders at perimeter areas and also nailed to the mudsill.

Cutouts were left in the perimeter girders and blocking to accommodate foundation vents. The larger opening is for an exterior crawl hole.

Interior blocking between girders consists of 2 x 4s spaced 40 inches apart, with each member toe-nailed in three places. Here the workman is using a pipe vise to force a slightly twisted girder into proper position for nailing.

The completed floor framing is being tested here for level and may be adjusted by moving the lock nuts on the adjustable jacks. Building department inspection is required at this point, and it is also recommended that necessary heating ducts and rough plumbing be installed before application of the subflooring.

A long-handled spade proved the most useful tool in removal of the garage roof at the major remodeling. The tar-and-gravel surface was scraped from the deck and loaded into a waste bin.

Decking was pried loose with a wrecking bar and hammer, and removed from the original framing members in preparation for new framing for the second-story bedroom.

Tongue-and-groove ⅝-inch plywood was used for the decking of the new master bedroom suite. Panels were nailed with staggered joints.

The old garage roof framing was 2 x 10 on 24-inch centers, insufficient for supporting the new bedroom addition, and thus removed.

The stucco garage walls were notched for placement of new floor joists installed on 16-inch centers.

New 2 x 12 floor joists for the master bedroom suite span the width of the two-car garage and create an overhang. The joists rest upon the original double 2 x 4 plate, plus an additional 2 x 4 plate atop the 2 x 4 stud walls. Blocking of 2 x 12 material was applied between joists.

Walls: Framing and Exterior Surfaces

Wall Framing

In all but the garage conversion, the newly built floors (and concrete slab) were used as the bases for assembly of the new wall sections, which then were lifted into place, temporarily secured in upright positions, and permanently nailed to adjoining new wall sections.

Standard 2 x 4 studs were spaced 16 inches on center and nailed to 2 x 4 plates. Window openings were framed at floor level, and the entire assembly then raised into position.

Walls for the bedroom-bathroom addition were started by temporarily nailing top and bottom plates together, placing them at the wall perimeter, and marking each for the location of studs and headers. Holding nails were then removed from the plates

and studs placed and nailed at the markings. An additional 2 x 4 plate and 1 x 4 plate were required to increase the wall height to that of the existing walls. This later required added gypsum wallboard beyond the standard 8 foot height.

These walls were also sheathed on the exterior side with ⅜-inch plywood before being lifted into position with a wall jack. Nailing to the subfloor was 2 inches on center. Shorter and lighter walls at the ends of the room were lifted into place by two workmen, without the aid of the wall jack.

Interior wall partitions were assembled in the same manner and nailed in place before start of the roof assembly, or before removal of the existing roofing where the tie-in was accomplished between new and old.

With the plywood platform completed for the master bedroom-bathroom home addition, workmen began assembly of exterior walls. Here the joined top and bottom plates are set along the front perimeter wall to be marked for window and stud locations.

The stud wall assembly includes headers at bath and bedroom window openings. Exterior plywood sheathing also was nailed in place before the wall was lifted into position.

Marking both top and bottom members at the same time eliminates possible error in stud placement.

The two plates are then separated and spread, and studs are nailed in place at marked intervals.

Wall jacks were temporarily secured to the plywood deck to ease lifting the frame wall into place. Each workman cranked a handle to raise the wall without strain.

This same procedure was followed for the construction of walls for the second-story bedroom, with the addition of 1 x 4 let-in bracing and the elimination of the extra plates. Walls assembled for the lower-level family room involved angle treatments to accommodate large window and door areas detailed in the floor plan. Plywood sheathing was not applied until the walls were in upright position and fully secured to one another.

Exterior Wall Materials

Wood siding, plywood paneling, and stucco were the three basic building materials used for the exterior wall surfaces of these home additions. The garage conversion and accompanying carport were surfaced with redwood siding to match other exterior walls of the home. The other two home additions relied heavily on stucco for the exteriors, with plywood panels used to add interest to the second-story add-on.

Stucco Stucco is widely accepted in the construction industry as a good exterior protector in hot or cold climates and is noted for its fire protection. Some methods of the trade date as far back as 4,000 years.

Local stucco specialists involved in these home additions mixed the material at the sites in one-batch power units that handle 94 pounds of cement and 25 shovels of sand per batch. This, incidentally, is about the maximum one man can apply before running into consistency problems.

Preparation of the walls for stucco first involved the application of plywood sheathing, which then was covered with building paper and wire mesh. Metal trim was used at outside corners and weep screed along the lowermost edge of the stucco surface.

Two coats are involved in the normal stucco application, and a third can be added if texturing is desired. The first or "scratch" coat is applied to a ½-inch thickness, permitted to dry for approximately 7 days, and then surfaced with a finish coat that results in a total ¾-inch thickness.

Stucco mixed in the batch unit is transferred by wheelbarrow to a mortar board and applied to the surface with a 5 x 14-inch trowel. The first coat is "scratched" with a comb-like tool within 30 minutes of application to provide a proper binding surface for the finish coat.

Depending upon the area being surfaced, workmen usually space sections of scaffolding approximately 7 feet apart, apply the stucco to that width, and then "scratch" the section before moving farther along the wall.

Aside from waiting a week between applications, stucco experts strongly recommend that the finish

Stud walls are joined at the top with a 2 x 4 plate shown here. The framing is nailed to the subfloor at the plate and, vertically, along the connecting members.

The sheathed perimeter walls, when fully in place, were complete with the exception of the bathroom window framing left out of the opening shown in the foreground. This was done to maintain a convenient construction "entrance" for workmen and materials.

coat be applied after the interior drywall has been hung. This finish coat can be textured with a brush and water to give a swirl or other pattern to the surface when it is dry enough not to show a fingerprint.

In joining new stucco to existing stucco on a wall surface, the wire mesh from the original application should extend into the new area. Special care should be given that the first coat is not so thick at this point that it prevents a second coat, or causes the second coat to be thicker than the original.

Stucco also was used to enclose the roof rafters at both levels of the two-story home. Wire mesh and a continuous screen vent were first applied and then the mesh was covered by stucco to simplify future house painting. All painting of new stucco should be delayed until at least a month after application of the finish coat.

Interior Walls

Framing

On-the-job framing of interior load-bearing and nonload-bearing walls and partitions usually involves the use of 2 x 4s for the structural members. The finish surface may be gypsum wallboard, paneling, or plaster.

Garage Conversion The garage conversion required only the construction of a storage wall, two walls to frame the bathroom, and a continuous windowless wall along one side to form a 6 x 19-foot storage area accessible from the side yard. The framing of this later wall did include the roughing-

Because it was lighter in weight than the plywood-clad exterior walls, interior framing could be raised into place by a single workman. This partition is joined to the exterior wall with a 2 x 4 top plate and nailing at the block locations.

in of an interior door frame which, now concealed by paneling, may later be completed by removing a single panel.

All of the garage partition walls are nonload-bearing, and most exceed the typical 8-foot height since the original garage height was maintained and a center 4 x 12 beam was converted into a lighting trough.

Each of the partitions has 2 x 4s on 16-inch centers and is surfaced with ½-inch gypsum wallboard and 4 x 8 sheets of ¼-inch Philippine mahogany. The finish panels were matched as closely as possible at the joint line above the 8-foot height.

Master Bedroom-Bathroom Addition Interior walls for the master bedroom-bathroom were framed on the plywood platform foundation, following erection of the exterior walls. A chalk line was first used to snap the actual layout on the floor, and

Framing detail of the bathroom window section of the master bedroom-bathroom addition shows how additional top plates were applied to increase the new room height to match the home's existing wall height. Note also the use of cripples above the window header.

Interior stud walls were constructed of 2 x 4s with assembly taking place on the plywood subfloor.

then material was cut and laid out to exact dimensions.

Door headers were framed into the partition walls before the assembly was lifted into place, positioned along the chalk line, and nailed to the floor. With all partitions in place, top plates were applied to complete the wall framing.

Two-Story Expansion Similar framing methods were used to create partition walls for the major two-story home expansion, including new walls in the lower-floor bedrooms to replace outdated storage cabinet room dividers.

As the existing storage cabinets were removed from each of the bedrooms, temporary support was provided to the roof joists. The new partitions were assembled, raised in place, and secured to the concrete slab floor using a power stud driver.

Finishing Materials

Gypsum wallboard was selected to finish off the walls in the bedroom-bathroom and the two-story home additions. This material is available in sheets 4 x 8 to 4 x 14 feet in size and in thicknesses of ¼, ¾, ½, and ⅝ inch. Three basic types were employed: standard tapered-edge, fire-rated wallboard (for the stairwell), and water-resistant for baths and kitchen.

Before any drywall could be applied, all interior framing, wiring, plumbing, and ductwork that was to be concealed had to be inspected by the local building department.

Application Drywall is first applied to the ceiling and then to the walls, placing wall units horizontally and beginning with the top sheet nearest the ceiling.

The additional height (beyond the normal 8 feet) of these walls required an extra 2-inch-wide strip of gypsum wallboard, which was applied between two 4-foot-wide sheets. This placement at midwall-height simplifies taping, as opposed to adding the narrow strip at either the top or bottom of the wall.

Standard drywall nails were used to attach the ½-inch-thick panels to ceiling joists and wall studs. A keyhole saw and circular cutters were used to notch or cut holes for plumbing lines and electrical receptacles.

Taping With all the drywall panels nailed in place, a contractor's bazooka tape gun was used to tape the joints. This tool speeds placement of tape and bonding material in the angled areas of walls and ceilings, and also is used for flat surfaces where panels are joined.

During warm weather, one day is usually sufficient for allowing the first application of taping to

Nonbearing storage walls dividing the first-floor bedrooms of the major remodeling project were removed to make way for new stud wall construction. The face frames were salvaged for other use.

New stud walls were nailed together at floor level and then secured to the ceiling and exterior wall, which was stripped of wallboard.

With ceiling wallboard panels in place, workmen applied all panels working from the top down. The wall shown here formerly was the exterior wall of the house.

An extra section of drywall was required in the center of the wall area to accommodate the 8-foot 3-inch ceiling height required to keep the new home addition in line with the original house construction.

A professional taping tool is used here to "spot" nailheads with a single downward pull.

A wider taping tool is used for the second coat application over horizontal panel joints.

Cut to fit, the gypsum wallboard was "pre-nailed" at lower level, raised into position, and finish-nailed as shown here.

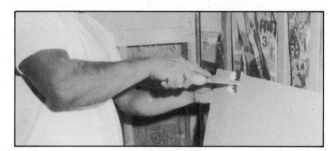

A keyhole saw is used to notch gypsum wallboard for fitting around pipes or electrical receptacles. A circular cutter also is used.

Equipped with an angle applicator, this tool was used to apply the second coat of taping mud to corner joints.

thoroughly dry. In damp or cold weather, several days may be required. In the major remodeling program, a portable heater-blower was used to shorten the drying process.

After all ceiling, corner, and flat joints were taped, the same all-purpose drywall "mud" was used to "spot" cover nailheads. A 6-inch-wide drywall knife was used for this purpose and also was used around windows.

The second taping step, called skimming, involves the coating of all seams using a 10-inch knife on flat surfaces and an 8-inch knife for outside corners, window surrounds, and nailheads. This application again requires a day or more for drying before the final texturing coat may be applied. When dry, a light sanding is done to remove any high spots.

Texturing Texturing of gypsum wallboard involves the same all-purpose taping mud, only mixed to a looser consistency by adding water and then adding fine sand to the mix. In this application, two small handfuls of Del Monte No. 30 sand were added to 5 gallons of mud.

An 18-inch knife with long or short handle is used to apply the texture coat. This skip troweling was done in a swirl pattern where walls were to be painted and was omitted where surfaces were to be finished with wallcoverings.

In working with drywall, it is well to keep the area

as clean as possible. Cutting and applying gypsum wallboard creates considerable fine white dust, which should be removed before taping applications begin. A nose-mouth mask is recommended, especially during the sanding phase.

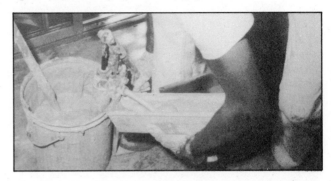

A hand pump simplifies removal of drywall mud from a 5-gallon container to a hand tray. The pole is used for frequent mixing to maintain desired consistency.

Finish texturing is accomplished by skip troweling in a swirl pattern.

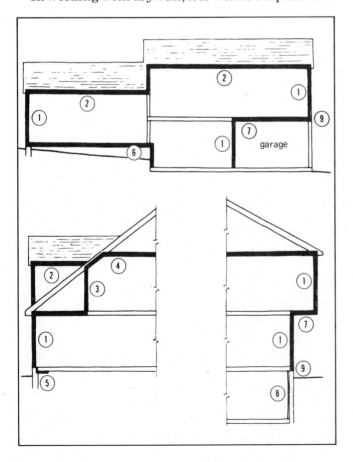

Insulation

Fiberglass batt insulation with its foil vapor barrier facing the warm (inside) side of the room was used to insulate all three home additions. Check the accompanying chart to see how much insulation you need in your part of the country.

Specific areas which require insulation are: (1) exterior walls and walls between unheated and heated parts of the house; (2) ceilings with cold spaces above, and dormer ceilings; (3) knee walls when attic is finished; (4) between collar beams and rafters; (5) around perimeter of slab; (6) floors above vented crawl spaces; (7) floors above unheated or open spaces, garage or porch; (8) basement walls when space is finished; (9) in back of band or header joists. Drawings courtesy of the Mineral Wool Insulation Assn. Inc.

Type of Insulation					
	R-11	**R-19**	**R-22**	**R-30**	**R-38**
Batts or Blankets					
Glass fiber	3½-4″	6-6½″	6½″	9½-10½″*	12-13″*
Rock wool	3″	5¼″	6″	9″*	10½″*
Loose Fill (Poured-in)					
Glass fiber	5″	8-9″	10″	13-14″	17-18″
Rock wool	4″	6-7″	7-8″	10-11″	13-14″
Cellulosic fiber	3″	5″	6″	8″	10-11″

*two batts or blankets required.

The following color section has been provided to stimulate ideas and designs for your addition and to show the types of materials and products available. The most popular projects are presented: bedrooms, bathrooms, family rooms, and kitchens.

Elimination of a window in the living room of the major remodeling was accomplished by installing studs from header to sill and insulating the area with fiberglass batts. A plywood panel was placed over this exterior surface, which was then stuccoed. The interior opening was closed off with gypsum wallboard.

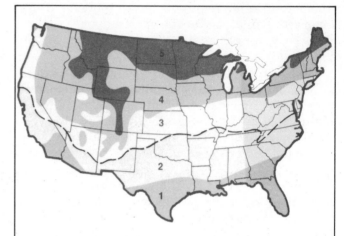

ATTIC INSULATION FOR WINTER HEATING								
HEATING COST				RECOMMENDED INSULATION				
GAS (therm)	OIL (gallon)	ELECTRIC RESISTANCE (kWh)	ELECTRIC HEAT PUMP (kWh)	ZONE 1	ZONE 2	ZONE 3	ZONE 4	ZONE 5
9¢	13¢	—	1¢	—	—	R-11	R-11	R-19
12¢	17¢	—	1 3¢	—	—	R-11	R-19	R-19
15¢	21¢	—	1 7¢	—	R-11	R-19	R-30	R-30
18¢	25¢	1¢	2¢	—	R-11	R-19	R-30	R-30
24¢	34¢	1 3¢	2 6¢	R-11	R-19	R-30	R-33	R-38
30¢	42¢	1 6¢	3 3¢	R-11	R-19	R-30	R-33	R-38
36¢	50¢	2¢	4¢	R-11	R-30	R-33	R-38	R-44
54¢	75¢	3¢	6¢	R-11	R-30	R-38	R-49	R-49
72¢	$1 00	4¢	8¢	R-19	R-38	R-44	R-49	R-60
90¢	$1 25	5¢	10¢	R-19	R-38	R-49	R-57	R-66

Find the amount of insulation you need on these weather maps and tables. Tables and maps courtesy of Certain Teed Products Corp.

Splines of galvanized metal containing top and bottom air holes were used to provide sound control at the exterior wall of the new bedroom-bathroom addition. As shown here, the metal channels were applied directly to the studs after the installation of batt insulation. Gypsum wallboard then was screw-attached to the metal, leaving an air pocket between the insulation and studs and the wallboard.

Fiberglass batt insulation has foil vapor barrier facing the warm side of the room. The material was easily applied with staple gun between wall studs.

This master bathroom features durable, easy-to-clean ceramic tile and exposed wooden beams. The whirlpool is built on a platform covered by tile. Photo courtesy of Eljer Plumbingware

This addition was fully insulated between rafters and on outside walls, then covered with paneling. A series of platforms were built to eliminate need for furniture and to provide bed, work area, dining area, seating, and storage. Photo courtesy of Champion Building Products

This simple addition involved recessing the toilet and towel storage area into an existing wall. A window was included to permit more natural light to enter the room. Photo courtesy of Eljer Plumbingware

Traditional fixtures in a contemporary setting create a striking bathroom. Cedar planks and plant groupings give a natural appearance and feel to the added bathroom. Pedestal sink and footed tub fit in nicely with the style of the home. Photo courtesy of Hedrich-Blessing

An addition like this leaves very little to the imagination. The master bedroom in the background opens through sliding glass doors into this elegant bathroom. Photo courtesy of Eljer Plumbingware

Wallpaper, carpeting, and molding converted this addition into an attractively furnished contemporary bedroom. Photo courtesy of Hedrich-Blessing

Most bathroom additions are larger and more elaborate than existing bathrooms. Tile makes this whirlpool tub very attractive and functional. Photo courtesy of American Olean Tile

This handsome bedroom is adjacent to a beautiful patio. Access is through the sliding doors. Photo courtesy of Georgia Pacific

This master bedroom suite is a simple addition. The curtain extending the height of the wall behind the bed gives the appearance of a window. Photo courtesy of Masonite

A fireplace was included in this bedroom-bathroom-dressing area addition. Notice the log walls in the background, part of the original home's exterior walls. Photo courtesy of Eljer Plumbingware

This bright, attractive family room addition takes advantage of a southern exposure to permit plenty of light to enter, during the day, through the floor to ceiling windows. Photo courtesy of Congoleum

Additions don't have to be square or rectangular. The angular design of this room creates a charming entertainment area. Photo courtesy of Armstrong Cork Co.

Here is another example of a major remodeling project featuring fireplace, loft, and kitchen. Photo courtesy of Georgia Pacific

This beautiful sunroom was added to bring the outdoors inside. Tile, tongue and groove ceiling and walls, and exposed beams add natural charm to the room. Photo courtesy of Martin Industries

This first-floor addition is finished in natural wood with bamboo squares on the ceiling and knotty pine that requires very little maintenance. The fireplace corner is real brick. Photo courtesy of Congoleum

In this addition (above left), planking was installed on the diagonal around the fireplace to create a striking focal point. Photo courtesy of Vermont Weatherboard

Masonry and paneling combine nicely in this bedroom-family room addition. Photo courtesy of Martin Industries

Planking was used in this family room addition. The existing dining room was also remodeled with the same material. Photo courtesy of Georgia Pacific

Open design is popular in dining room-kitchen additions. This design creates a beautiful work and entertainment environment. Photo courtesy of Congoleum

The wall between family room and kitchen was opened up to create an informal dining table. Photo courtesy of Wilsonart

This kitchen-dining room addition was simple to build and offers plenty of good working area. Photo courtesy of Wilsonart

This open area functions as kitchen, dining room, and family room. The two levels are unified by the same floor covering and color scheme. Photo courtesy of Armstrong Cork Co.

The New Roof

Extending the typical home upward or outward to obtain more living space almost always requires cutting into the existing roof structure. Here again, the scope of the activity will vary with each given plan, ranging from a complete removal of the roof to add a second story, to the removal of a small area into which the new addition's roof will be tied.

Both the master bedroom-bathroom addition and the major remodeling fell into this category, with the blending together of new and old taking place at the ridges of the homes. The second-story add-on brought about an entire roof change for the home, as the owners switched from composition shingles to medium-weight wood shakes. The bedroom-bathroom addition involved blending of new, heavy shakes with the existing ones.

Design and Preparation

Proper design of an expanded roof is extremely important to the success of the overall home addition. The structure, of course, must meet code regulations, but it should also appear to be a part of the original home construction.

The pitch of the new roof should correspond to the pitch of the existing roof line and materials should match as well as physically possible. Use of two different roofing materials will only highlight the fact that you have added on and can negatively affect the sale of the home in later years.

The act of joining a new roof area to the existing structure begins with the removal of gutters, overhang, and roofing materials at the points of connec-

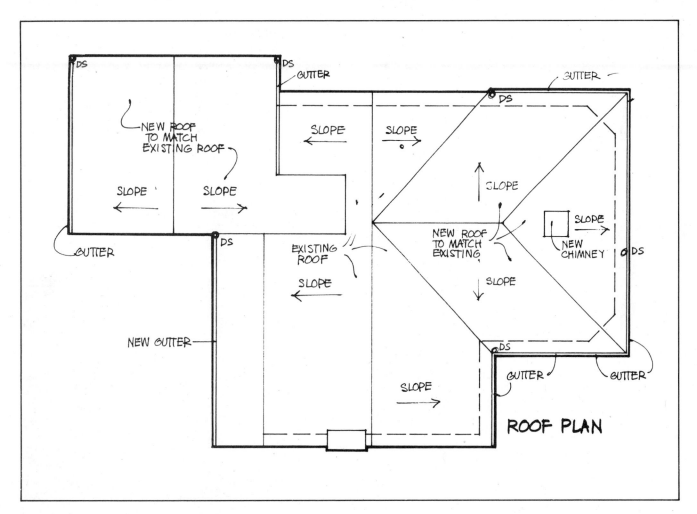

This architectural drawing indicates the various slopes required for new roof areas at the front and rear of the major home expansion, and details the locations of gutters and downspouts.

Standard 2 x 8 roof joists, used on both home additions, were cut at ground level and nailed atop supporting walls on 16-inch centers.

The roof joists span the entire depth of the second-story addition, running from the front of the garage to the rear of the new addition.

tion. A power saber saw is most useful in accomplishing this.

In planning the first-floor master bedroom-bathroom addition, the remodelers designed a hip roof with 4/12 pitch. The ridge of the new structure comes together at a 45° angle to the ridge of the existing structure.

Framing

Physical layout of the framing began with placement of the new ridgeboard from the inner wall of the addition to the existing roof ridge. This was temporarily held in place with wood braces while the workman cut and attached the 2 x 10 common rafters which formed the basic structure. Intermediate 2 x 6 rafters then were cut to length and nailed to the ridge and common rafters, allowing an 18-inch overhang on three sides of the addition.

Once the framing was completed above the new rooms, and a ridge was built to tie the assembly to the existing ridge, 1 x 6s were nailed to the roof and additional rafters secured above the existing home. At a later date, the owner decided to remove the existing wood shingles which otherwise would have been below the new roof assembly.

With all of the rafters in place, 2 x 4 frieze board was applied at the building perimeter, between rafters and in combination with screen vents, to ventilate the new roof area. Three rows of tongue-and-groove 1 x 8 boards were applied atop the rafters at the perimeter to form the overhang.

The balance of the roof framing consisted of the nailing of 1 x 4 boards (with an equal space between each) to the rafters to accommodate nailing of wood shakes applied over common building paper.

New ridgeboard and 2 x 10 common rafters extending from the existing roof of the master bedroom-bathroom addition were temporarily supported by 2 x 4s attached to the roof and new roof framing.

Working from the roof joist, the carpenter next secured the new hip member to the ridge assembly.

Roof rafters of 2 x 6 material then were cut to proper pitch and length and nailed to the common rafters on 24-inch centers.

With rafters applied the entire length of the common rafter, members then were added to the other side of the 2 x 10.

This view of the new roof framing shows temporary support in place for the new ridge, rafters applied over the new addition, and the framed attic hatch in the ceiling joist. Note how the joists are joined at the center supporting wall, a factor that made it possible to use shorter material for this assembly.

After all new rafters were applied over the new addition, rafters were then applied from the ridge to 1 x 6s nailed to the existing roof. Shakes were not removed at this time, but later were removed from inside the new roof structure as a fire safety precaution.

Following the nailing of 2 x 4 blocking between rafters at the perimeter walls and placement of roof ventilation screens in appropriate openings, tongue-and-groove frieze boards were applied working from the rafter ends toward the ridge.

Roof sheathing next was nailed to rafters. As shown here, over a dozen 1 x 4s were stacked atop the rafters, with the workman nailing every other one in place. The loose boards served as exact spacers and then were moved to higher locations.

Valley flashing was installed where the new roof meets the existing one, with the metal secured under the felt layers.

Roll gutters matching those of the existing structure were nailed to the frieze board and were further supported by galvanized metal straps running from the front edge of the gutter to the frieze boards.

Soldering irons, used to fabricate sheet metal at the bedroom-bathroom home addition, were heated in a butane fire pot. Shown here are the sections of the skylight flashing. Straps at right are gutter supports.

With all sheathing in place, the sheet-metal worker fabricated flashing for the roof skylight located above the bathroom-dressing area.

Flashings and Gutters

Standard roof flashings were used in both remodeling projects involving new roof structures. The galvanized valleys and slip-over pipe flashings were nailed in place as the shake applicators progressed. Gutters were installed by a sheet-metal contractor prior to the start of the shake application.

While most flashing items required for home additions are factory-made and merely require proper placement and securing on the job, gutters must be fabricated at the job-site to meet specific length and angle requirements. In two of the remodeling projects, special forming was required to match the existing gutters, which were not being replaced.

Both left- and right-handed aviation snips are used by the sheet-metal worker to cut and shape gutters. Once the gutter is cut, muriatic acid is applied to the soldering surface to clean the metal of grease and assure a better bond.

A small butane fire pot is used to heat soldering irons—usually 4-lb. units sold in pairs so that one is being heated as the other is being used. The gutter sections require a 1-inch lap at joints and are attached to the roof with special metal hangers spaced every two or three rafters.

Downspouts also generally require some on-site fabrication to provide a proper angle from gutter-to-house and house-to-yard drainage. Metal straps hold these units in place.

Aviation snips were used to cut sheet-metal items, including this downspout which was notched for bending to the proper shape.

Some sheet-metal joints had to be soldered at roof level, such as this L-shaped fitting of two new gutter sections. This view is before soldering.

Applying Wood Shakes

In applying wood shakes, workmen used 18-inch-wide rolls of 30-pound felt for underlayment, which provides one course under every row of shakes. The felt is started at the eave and each layer overlapped toward the ridge. Standard roofing nails are used to secure the paper to each rafter, and care is taken at the eaves not to nail through overhang boards and leave a visible point.

The felt extends across the valley flashing but is kept approximately 2 inches from the crown (shakes are kept 1½ inches from the crown).

Wood shakes are sold in "squares" with each square consisting of five packs or enough shakes to cover 20 square feet of roof surface. When possible, the materials are lifted directly to the roof via scissors-bed delivery truck, eliminating considerable handling and use of a ladder.

As each bundle is opened, workmen select shakes that are a foot or greater in width and use these for valley areas. A power saw usually is used to cut angles required along the valley. Another trick of

Notched and bent to the desired shape, the new downspout joint was soldered to produce a watertight seal.

With wall and roof framing in place at the major home addition, a plastic blanket was applied over the entire assembly to protect open areas of the existing home roof and the interior from rain. The plastic sheeting was removed when the shake roof was applied.

Beginning at the eave edge of the roof sheathing, a workman applies 18-inch-wide felt, overlapping each course and spot nailing to hold in position.

With all of the 30-pound felt in place, stack and vent flashing is positioned for plumbing columns.

the trade is to use the wood-strapping band on each bundle as a measuring device in holding all shakes an equal distance back from the flashing crown.

A special starter course is used at the eave, with No. 6 nails being driven in part way, and then folded down to prevent full penetration of the overhang boards. This practice is used for the first three courses of shakes.

Workmen limit the amount of moving they must do on the roof by applying shakes for six rows in an area approximately six feet wide, and then they move to the next location. Shakes should have a maximum weather exposure of 10 inches for 24-inch units.

A ½-inch space between individual shakes is left to allow for possible expansion. These joints or spaces should not occur directly above similar joints in a straight line.

Special shakes are available for capping hips and ridges and are applied with 8d nails.

Larger-width shakes are selected from each package as it is opened and are used along the metal valley. A power circular saw was used on this job to cut the required valley angles.

Starter shingles used in the first course along the eave line are doubled. Each succeeding course is applied by tucking the thin edge under the felt above.

Cap shakes assembled at the factory are used to finish hips and ridges.

Blending new shakes into the existing roof structure involves application of new roofing felt a foot or more over the old felt below the shakes and above the flashing. New shakes are then applied.

Extending the Utilities

The extension of water supply, sewer line, heating ductwork, and electrical service from the existing structure to an addition will vary greatly in degree of difficulty. Fortunately, many homes built over the past twenty years have an amount of unused "service" which can be tapped for the new addition.

The simplest extension covered in this book was for the garage conversion. It involved: tapping into the existing gas line to supply a new 17,500-BTU wall heater, which operates independently from the balance of the home heating system; pulling two more electrical circuits from the existing panel source (one circuit already existed in the garage for normal lighting and utility use); and, tying into existing plumbing lines which backed up to the new bathroom.

For the master bedroom-bathroom addition, the "rose garden" addition was complicated, in that a new electrical panel box was required to bring the existing electrical wiring up to code, new ductwork was extended in the new crawl space to provide heating, and plumbing lines were extended from existing lines located in the old crawl area, across the bedroom hallway.

The major remodeling required the most work, for individual gas-fired wall heaters throughout the house were removed and replaced with a central gas-fired unit (on the second floor) and new ductwork. The original electrical panel box was moved from the former exterior rear wall to the side wall of the new family room (it was large enough to handle additional new circuits), and new bathroom plumbing lines were connected at the utility location in the garage and nearby sewer line. All of the original wiring in the kitchen was replaced.

Plumbing Service

Master Bedroom-Bathroom Normal construction practice calls for the installation of new plumbing in a crawl space before the placement of subflooring on the framework. However, in the bedroom-bathroom addition, the plumber's work load elsewhere prevented him from getting to the job for several days, and the carpenters were permitted to apply the subflooring. A single plywood panel was left loose and then removed by the plumber, who worked on his back in the crawl space to make the necessary hookups to existing lines.

Tyler no-hub cast iron was used for the drain-

Cast-iron plumbing fittings used to extend waste and vent lines to the new bedroom-bathroom addition were joined to cast-iron pipe with neoprene gaskets and stainless steel bands, shown in the foreground.

waste-vent system, a factor that reduced installation time to approximately 25 percent of the time that would have been required with a galvanized system. The cast-iron pipe is easily reduced to needed length with a chain cutter that "snaps" the cut. Components of the cast-iron system include neoprene gaskets, which slide easily over the ends of pipes being joined together, and stainless steel bands to secure the couplings in place when tightened with a 60 lb. torque wrench.

Both 1½- and 2-inch diameter cast-iron pipes were used for venting purposes in the bathroom plumbing system. The supply lines were extended from existing galvanized lines, using dielectric unions to prevent electrolysis between the galvanized and the new rigid or flexible copper tubing.

Water lines for the toilet and bidet were installed approximately 6 inches above the floor line, while lavatory supply lines were installed for 8-inch center faucets. Standard soldering methods were used in making copper connections. For more information and details on pipe installation, please see *Homeowner's Guide to Plumbing.*

The 32-inch-wide shower stall was fitted with dual heads, one on each end wall at a 6-foot height, and single-handle valve controls at 4-foot height. Felt was used between pipe and securing brackets to prevent possible rattling noise.

Installation of the rough plumbing—both supply and vent lines—involves capping the lines before actual fixtures are installed, filling the system with

A saber saw and circular hole saw were used by the plumber to cut through subflooring and wall framing to extend waste, vent, and supply lines.

Working to predetermined measurements, the plumber subcontractor assembled many of the new fittings outside of the crawl space area. Shown here is the waste line clean-out end.

A small pocket was provided between the end shower wall and linen closet-to-be and used as the chase for vent stacks to the roof. Vents from the toilet, bidet, and shower were connected at this point to a single vent pipe penetrating the roof.

Soft copper lines were used here for the lavatory supply lines due to the layout of foundation framing. Once in place, caps were soldered on the ends so the entire system could be tested for leaks before fixtures were installed. Hot-and cold-water lines are attached to wood blocking, providing water for the bidet. An extra hole in the subfloor results from a switch of bidet models, the newer unit calling for a drain line farther from the wall.

The completed vent assembly shows the toilet vent coming from the left, bidet line at right, and shower line in the center, all joined to the single column leading to the roof. This eliminated the need for several roof jacks.

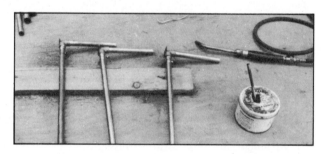

Rigid and flexible copper tubing was fabricated for water-supply lines at the shower, vanity-lavatory, water closet, and bidet. The ½- and ¾-inch lines also included a hose bib to the front wall. Shown in this photo are plumber's torch, flux and solder, and torque wrench and driver.

Completed lavatory plumbing assembly shows supply lines for the twin bowls, waste line for each (temporarily capped), and vent lines running to a common 2-inch vent line extended through the roof. The copper assembly at left includes a ¾-inch cold-water line, off of which is the outside hose bib.

water drawn from a hose or from tapping into the existing source, and testing for leaks. In most areas, city inspectors must approve and "sign off" this phase of the installation before the plumbing is continued.

It is well to have specific model fixtures and control fittings designated when plumbing extension is involved. Doing so will enable the plumber to "plumb" for the given unit, cutting necessary holes in the subfloor and placing lines where they will be needed. Bidets, for example, require waste lines at various distances from a wall, depending upon the brand and model number selected.

Second-Story Major Remodeling Extension of the supply and waste lines for the new second-story master bedroom suite conveniently originated from the interior rear garage wall, which was the location of both the hot-water tank and automatic laundry units. This area also was originally equipped with a cast-iron laundry tub, which was removed and eliminated in the remodeling.

By cutting into the supply lines serving the automatic washer and dryer, the plumber was able to "tee" the new copper lines, run them across the ceiling joist and upward to the second floor via wall framing. Supply lines were required for shower, water closet, and single-bowl lavatory.

The concrete garage floor had to be opened to tie in the waste line which runs to this point of the house, and then outward to the street sewer line. Cast-iron pipe was used for both waste and vent lines.

The completed vent line for the lavatory is pictured in the foreground along with copper supply lines. The unfinished stack in the background is in the toilet compartment and was extended through the roof as an independent vent.

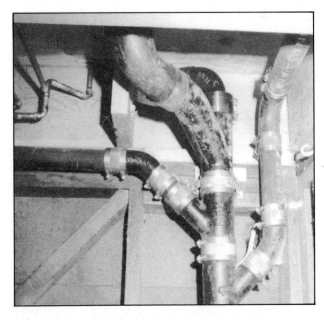

The upper area of the waste connection in the garage shows lines leading off to the lavatory, water closet, and shower stall. The copper tubing, at left, is the water supply, which was extended from laundry utility lines in the garage.

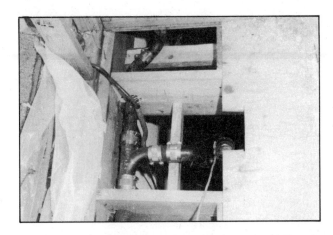

Viewed downward through the second-story subfloor is the waste assembly for the new shower stall. The electrical cable for telephone service was rerouted.

New vent pipe was required as the gas-fired water heater was moved and enclosed in a corner of the garage. Walls of the enclosure are surfaced with noncombustible panels, and a screened vent at the base of the wall provides fresh air from the front entryway of the house.

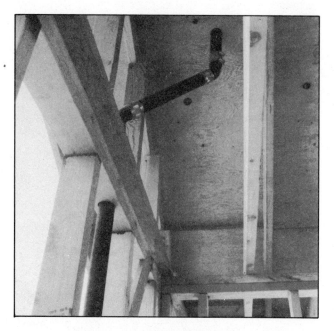

Cast-iron vent lines run through the wall framing and are then angled to penetrate roof sheathing, placing the roof vent several feet inside the roof edge. The second-story walls are shown here before sheathing was applied.

Heating

Master Bedroom-Bathroom The heating extension for the bedroom-bathroom addition was a relatively simple one, involving flexible ductwork run to the existing heating unit.

Extending heating to the new bedroom-bathroom was a relatively simple project. Flexible ductwork was used for runs from the existing heating unit, through the new crawl space wall opening (covered here by plywood) and to floor register locations.

It was from this garage-utility area plumbing that supply lines were connected and extended to the second floor. A change was made in the waste line, eliminating the previous clean-out and keeping all of the lines within the wall framing, which later was covered with gypsum wallboard.

Galvanized straps were used to secure the insulated ductwork to floor girders. The ductwork was delivered to the site completely wrapped in vinyl-fiberglass.

Lock-together fittings joined ductwork with the floor register; the insulation was pulled across the connection and taped in place.

Second-Story Major Remodeling The new central gas-fired unit for the major home addition involved putting in a totally new heating system. Although a considerable investment, it will pay off in the long run because of the home's higher resale value and increased comfort.

The gas-fired, warm-air furnace, used to replace wall heaters in the major home addition, was raised to the second level of the home before wall framing was completed. The unit is located in a compact furnace room at the top of the stairs, outside the new master bedroom suite.

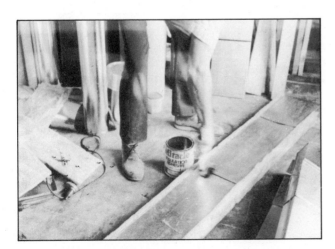

Snap-together ductwork used throughout the house was first insulated at the job-site. Fiberglass was applied to the metal with adhesive.

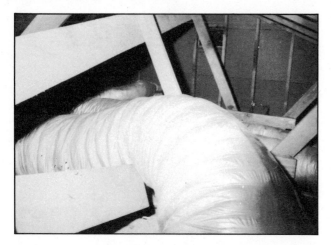

Flexible, insulated ducts used to move warm air (and cold air return) were secured to the heating unit and pulled through the attic area to specified locations.

Electrical Work

Electrical service for a home addition varies, of course, depending upon the activities to take place in the addition. Adding a bedroom or study that requires only lighting and convenience outlets is much less complicated than adding a new kitchen or laundry, which would require lighting, convenience outlets, and a power source for major and minor appliances.

Master Bedroom-Bathroom The bedroom-bathroom addition required replacement of the main service panel at the rear of the home, which was a

The new Square D service panel was installed horizontally below the electric meter. Existing house wiring was connected to automatic circuit breakers, and a new cable line (left of the meter) was extended to a subservice panel installed in the wardrobe closet of the new addition.

Designed to fit standard stud spacings, the ductwork was installed in openings previously used for wall heaters and was also installed in the new rooms.

thirty-year-old fuse panel with two 30-amp sub-feeds for the outdated knob-and-tube wiring used throughout the dwelling.

In replacing the service panel, a Square D load center was located beneath the electrical meter box, and existing house lines were connected to individual breakers. A 6-12 line was then pulled from the new panel to a new master-bedroom wardrobe closet. This arrangement eliminated the need to pull each separate new circuit into the home addition throughout the existing house and back to the central source. Wires instead were pulled from the wardrobe closet box to lighting locations, convenience outlets, and the sauna controls.

It should be noted at this point that the non-metalic or Romex cables used for all of the home additions detailed in this book were grounded. This means the neutral conductor (white or gray color) is connected to a single system ground and runs through the circuitry back through the distribution panel, and perhaps to the service entrance. Wire sizes are designated by number; the lower the number, the greater the current capacity.

The No. 8/3 cable pulled through the attic to the new bedroom panel box served as the source for separate No. 8 lines to the sauna, No. 12 line to a

Spools of nonmetallic cable (Romex) were hung from the ceiling on a dispensing reel, which speeds pulling the material throughout the area being wired.

A power drill with extra-long bit was used to drill holes in top plates and studs.

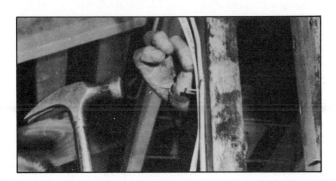

Staples were used to secure Romex to framing members.

Each cable end was stripped as required to electrify switches, receptacles, and fixtures as well as to provide grounding.

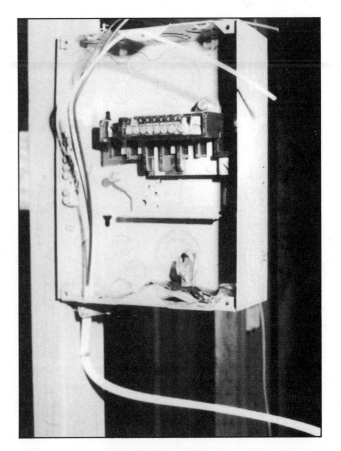

Each circuit is pulled through "knock-out" holes in the service panel and connected to the distribution "bus." The main supply line from the main distribution panel is the black cable which enters the right, rear area of the box. The light-colored lines are new circuits for the home addition.

Special framing sometimes is required to accommodate lighting fixtures such as this one for combination heat-light used in the bathroom. It is supported by 2 x 4s nailed to ceiling joist.

Adjustable placement of ceiling fixtures is possible within joist spacing, through use of bar hangers shown here.

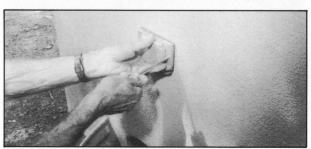

Ground-fault-interrupter (GFI) outlets were installed at two locations in the exterior wall of the bedroom-bathroom addition, to be used for future outdoor lighting and power tools. The units immediately cut the circuit power in case of overload or other dangerous situations. The outlets have weatherproof covers.

bathroom heater, and No. 14 lines to lighting and convenience outlets.

This installation required two standard circuits, plus a third for the sauna. Depending upon the area code, some cities permit lighting fixtures and convenience outlets to be wired on the same circuit, while other cities require they be separated.

Second-Story Major Remodeling Wiring for the major home expansion project resulted in virtually rewiring the entire house. The original Zinsco 100-amp service was moved from the rear wall of the house to a side wall of the new family room, and six new breakers were added to the box to handle the added wiring requirements.

In making the service box relocation, a new roof jack was installed to handle the No. 4 line running from utility pole to the house; the old roof jack had been bent. Rigid conduit then was installed from the roof jack to the service box, as required by code.

The expanded electrical service box now supplies power for 20 circuits required throughout the house, including the following:

- A kitchen range No. 6 cable supplying 220 volts and utilizing two 40-amp breakers at the box.
- A No. 12 wire circuit for an automatic dishwasher and food waste disposer, located under the sink and with an above-counter switch for disposer operation. The dishwasher is merely plugged into this convenience box.

- Two No. 12/3 convenience outlet circuits for the refrigerator and other small appliances used in the kitchen.
- A No. 14 wire circuit for lighting.
- Two No. 12 wire circuits connected to a ground fault interrupter and used for outdoor convenience outlets and lighting.
- A No. 12 wire circuit for the new gas warm-air furnace.
- A No. 14 circuit for the convenience outlets in the new family room.
- A No. 12 wire circuit for outlets in the new master bedroom suite.
- Two circuits for a 220-30 amp No. 10 cable to the automatic dryer in the garage.
- A No. 12 wire circuit to the automatic washer.
- Six other circuits for the balance of convenience outlets and lighting throughout the house.

For Safety's Sake In order to meet utility company standards and local code requirements, as well as assure the safety of the people in your household, it is best that a licensed contractor install and connect:

- Service head mounting pole or bracket, or enclosure and meter base;
- Conduit or through-wall cable to the main disconnect-switch-and-fuse box;
- Conduit or cable through interior from the main switch to the distribution panel;

- Conduit or cable from the distribution panel serving as feeder to any branch distribution center;
- Provision for and connection to system ground.

This leaves only the more time-consuming but easier work of installing outlet receptacles, switches, and fixtures for the do-it-yourselfer. Provided here is a typical general purpose circuit to help you familiarize yourself with the terms and types of layout, plus information on outlet boxes, and charts on wiring materials and electrical demand. You can use these charts for figuring out your own needs and for understanding the many different factors involved in electrical projects for home additions.

Pictured here in its original location, the service entry panel for the major remodeling was totally disconnected and moved to a new location on the side wall of the new family room. The electrician is shown here connecting circuits to individual breakers.

Allowable Wiring in Electrical Boxes				
	Maximum Number of Conductors			
Box Size (inches)	No. 14	No. 12	No. 10	No. 8
1¼ x 3¼ octagonal	5	5	4	0
1½ x 4 octagonal	8	7	6	5
1½ x 4 square	11	9	7	5
1½ x 4¹¹/₁₆ square	16	12	10	8
2⅛ x 4¹¹/₁₆ square	20	16	12	10
1¾ x 2¾ x 2	5	4	4	4
1¾ x 2¾ x 2½	6	6	5	4
1¾ x 2¾ x 3	7	7	6	4

Note: Shallow boxes, less than 1½ inches deep should not be used in new construction.

Box Capacities for Wiring depend upon conductor sizes. Chart indicates the maximum number of conductors allowed for each box size.

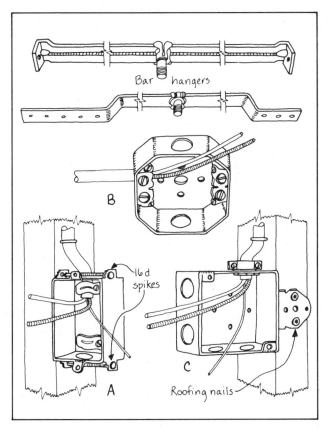

The most common types of outlet boxes used in homes are: (A) 1¾" x 2¾" on the face, and usually 2½- or 3-inches deep, with either side flaps or spike holes for nailing to studs; (B) a typical 4" x 4" octagonal box, 1½ inches deep, with a large center hole for direct mounting to bar hangers that nail to lower edges or to sides of ceiling joists; (C) a 4" x 4" square box, 1½ inches deep, with nailing side flap, used for convenience outlets. Installation devices that come with boxes either bring the front box edge flush with framing members or project inward ⅜ or ½ inch to penetrate wallboard.

Description—Typical Use Room Area/Sq. Ft.	Bath or Utility 50-60	Bedroom or TV 125-150	Bedroom, etc. 200-240	Family/Rec. 340-380
Wiring Materials Checklist for Typical Home				
No. 12/2 Conductor (with Ground) Safeguard Type NM Plastic Jacket	50 Feet	100 Feet	125 Feet	150 Feet
2-hole Wall Straps	use at 2-to-3 ft. intervals where cable needs support			
Steel or Plastic Boxes—With NM Cable Clamps and Ground Attachment				
Octagon or Round for Lighting Fixtures	1	2	2	2
Outlet Boxes for Duplex Receptacles	1	4	6	6
Switch Boxes—1 Gang or 2 Gang to match switches	1 or 2	1 or 2	1 or 2	1 or 2
3½″ Nails (if not provided with boxes)	4	10	14	14
Finish Wiring Devices				
15A-120V Single Pole Switch (If switching in 2 locations is desired use 3-Pole switches and 3 conductor grounded Safeguard Type NM)	1 or 2	1 2	1 2	1 2
OUTLETS—Duplex Parallel Slot 15A-125V	1	4	6	6
Switch Wall Plates, Single Pole	1 or 2	1 or 2	1 or 2	1 or 2
Duplex Wall Outlet Plates	1	4	6	6
Solderless Wire Connectors (for No. 12 Wire)	Minimum 6	6	8	12
1/2″ Type NM Cable Connectors	2	4	6	8

SPECIAL APPLIANCE WIRING CIRCUITS*

Electric Dryer or Electric Hot Water Heater	**Electric Range or Electric Furnace**
25′ or 50′ No. 10/3 Conductor (With Ground Wire Type NM or UF Plastic Jacket Cable)	25′ or 50′ 8/3 Conductor (With Ground Wire Type NM or UF Plastic Jacket Cable)
One—30 AMP Dryer Outlet	One—50 AMP Range Outlet
One ¾″ Cable Connector	One—¾″ Cable Connector
Six—2-Hole Cable Straps (One for 4-5′)	Six—2-Hole Cable Straps
One—30 AMP 220 Volt Circuit Breaker*	One—50 AMP 220 Volt Circuit Breaker*

*Maximum Conductor Capacity: 10/3—30 amps 8/3—40 amps 6/3—60 amps
SOURCE: Safeguard Electrical Products Corporation, from Safeguard House Wire in Pick-Me-Up Carton

Ckt. No.	Location	Lighting Outlets	Conven. Outlets	Estimated Watts	Amperes	Conductor No. & Size	Watts Load
Electrical Demand Factors and Load Calculation							
1-G	Dining, Laundry, Living	4	3	1250	15		
2-G	Entry, Kitchen, Bath No. 2	5	1	1250	15	3-#14	
3-G	Hall, Den, Bedroom No. 1	3	4	1100	15		
4-G	Bedroom No. 2, Bath No. 1	3	4	900	15	2-#14, 1-#12*	
	NOTE: General lighting circuits total 4500 watts or 3.00 watts/sq. ft.						
5-A	Kitchen, Dining		4	1000	20	2-#12	
6-A	Kitchen, Laundry		4	1000	20	2-#12**	
7-A	Dining, Laundry		3	750	20	2-#12	
8-A	Garage-Workshop		3	1250	20	2-#12	
				8500			
			First	3000	@ 100%		3000
			Balance:	5500	@ 35% (approx.)		2000
9-I	Range: 12kw model, max demand 8kw		1			2-#8, 1-#10	8000
10-I	Water Heater		1	4000	20	2-#10	
11-I	Bathroom Heater		1	1500	20	2-#12	
12-I	Furnace (1/4 & 1/200 HP motors)		1	600	20	2-#12	
13-I	Automatic Washer (1/6 HP motor)		1	450	20	2-#12	
14-I	Dishwasher		1	1000	20	2-#12	
				7550	@ 75% (approx.)		5700
							18,700

Calculated load for service size: Watts (18,700) divided by Volts (230) = Amperes (81.3) = 100-ampere service

*This #12 conductor is neutral common to circuit No. 11-I
**One of these #12 conductors is neutral common to circuit No. 13-I

Windows and Doors

During the design and construction of a home addition, older windows and doors often must be removed when the space is prepared for new living purposes.

In selecting windows and doors, attention should be given not only to style but to the finish that will be used. Many doors today are manufactured with "plugged" skins that show defects or patches when stained. These doors are designed for finishing with paint, which hides the patching.

Windows

Garage Conversion The original fixed sidewall garage window was removed in favor of an operable unit that would provide ventilation. The garage door was to be replaced by an aluminum sliding glass door to create a window wall.

The 16-foot opening for the garage door area was framed to include a 4-foot fixed, floor-to-header window on each side of a standard 8-foot sliding glass door. The entire area overlooks a private patio at the entry to the home.

Facing the street, the new garage window treatment consists of 3-foot-square wood sash, the center unit fixed in place with flanking units crank-operated for opening and closing. The new windows fit tightly into the original rough framing after the old aluminum unit was removed.

Master Bedroom-Bathroom In the bedroom-bathroom addition, two existing extra-high, double-hung windows were removed after the new structure was weathertight. New stud framing was added to seal the wall, using plaster on one side and new gypsum wallboard on the bedroom side.

Aluminum sliding windows were placed on the street-side of the new addition so as not to interfere with foot traffic through the garden. A 4 x 4-foot unit was used in the bathroom, and a high, 8 x 2-foot, horizontal, insulated (for sound control) model with two operating panels was used in the bedroom area. Both styles were set into the rough opening and nailed securely in place with metal clips provided by the window manufacturer.

Major Home Remodeling A variety of wood window styles was used to replace all existing windows in the major home addition, with dated steel units removed once the new windows were at the site. One living room window was totally eliminated from the fireplace wall, which was stripped of

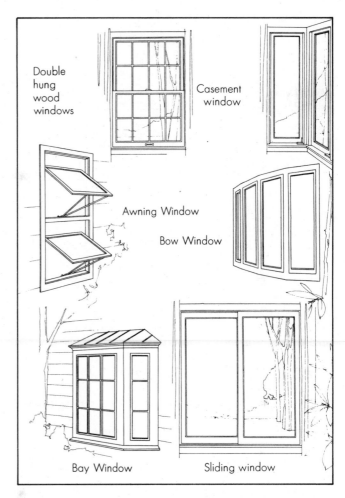

Drawings courtesy of Ponderosa Pine Woodwork

old wallboard, insulated, and covered once again with wallboard.

Many of the window openings were modified to make better use of wall area for furniture arrangement. In the bedrooms, new headers and additional under-window blocking were required as additional wall space was gained from windowsill to floor.

In other rooms, such as the living room and dining room, just the reverse occurred. Smaller windows were replaced with six-foot, six-inch casement windows used for the major wall areas of the family room add-on. All of the windows have inside-mounted screens and are operated by crank movement.

Replacement of the old windows required perimeter replacement of stucco, as part of surfacing the new exterior walls of the second-story rooms and the new family room.

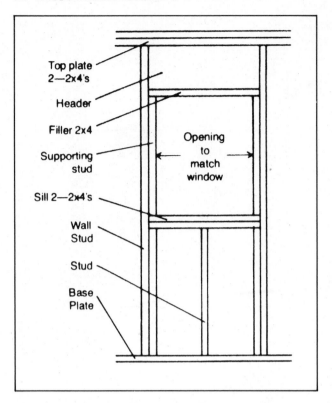

Top plate
2—2x4's

Header

Filler 2x4

Supporting
stud

Opening
to
match
window

Sill 2—2x4's

Wall
Stud

Stud

Base
Plate

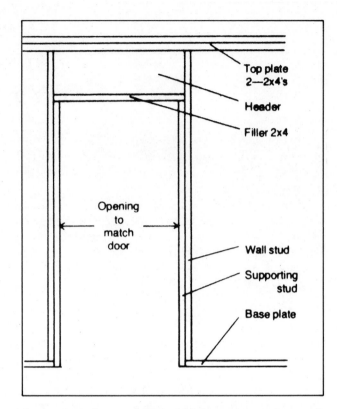

Top plate
2—2x4's

Header

Filler 2x4

Opening
to
match
door

Wall stud

Supporting
stud

Base plate

Preparation of the Rough Opening

Installation techniques, materials, and building codes vary according to area. Contact your local material supplier for specific recommendations.

The same rough opening preparation procedures are used for wood and Perma-Shield windows.

Brick veneer with a frame back-up wall is similar in construction to the frame wall in the following illustrations.

When enlarging the opening is necessary, make certain the proper sized header is used. (Contact your supplier for proper header size.) For installation of a smaller window—frame the opening as in new installation. Drawings courtesy of Andersen Corporation

1. *Lay out window opening width between regular studs to equal the window rough opening width plus the thickness of two regular studs.*

2. *Cut two pieces of window header material to equal the rough opening of window plus the thickness of two jack studs. Nail two header members together using adequate spacer so header thickness equals width of jack stud.*

3. *Position header at desired height between regular studs. Nail through the regular studs into header with nails to hold it in place until completing next step.*

4. *Cut jack studs to fit under the header for support. Nail jack studs to regular studs.*

5. *Measure rough opening height from bottom of header to top of rough sill. Cut 2 x 4-inch cripples and rough sill to proper length. Rough sill length is equal to rough opening width of window. Assemble by nailing rough sill into ends of cripples.*

6. *Fit rough sill and cripples between jack studs. Toe-nail cripples to bottom plate and rough sill to jack studs at sides.*

7. *Apply exterior sheathing (fiberboard, plywood, etc.) flush with the rough sill, header, and jack stud framing members.*

Before installation of the new wood windows, new framing was covered with plywood, building paper, and wire mesh for stucco. Kraft paper was also used around the perimeter of the window.

Replacement of steel windows throughout rooms in the major remodeling required new framing to create different sized openings. As shown here, 2 x 4 cripples were used to raise the sill height in bedrooms, and new headers were installed with standard 2 x 4 vertical framing techniques.

Exterior Doors

The aluminum sliding glass doors used for both the garage conversion and the bedroom-bathroom additions were set atop the plywood subfloor with a bead of caulking along the outside edge. Thus temporarily held in place, the door frames were leveled and plumbed, and nailed to the framing surround.

With the frame installed, the fixed panel section was put in first, followed by the sliding panel, according to instructions supplied by the door manufacturer. Installation of locking hardware completed the job.

New windows in the family room were set in place with a kraft paper surround, leveled, and plumbed, and then face nailed at predrilled locations.

The sliding glass door for the bedroom-bathroom addition was delivered to the site in semi-prefabricated form. The frame was fitted to the rough opening to check for fit, removed while a bead of sealant was applied to the subfloor, and reset in place.

The double-width overhead garage door removed in the garage conversion was replaced by this standard sliding glass door and flanking, floor-to-ceiling fixed glass windows. The existing garage door header remained in place, and new framing was installed below the new sliding glass door at the original garage floor level. Note the use of a redwood step, which serves as an entry landing between the new living-area floor height (level with the base of the sliding glass door) and the new aggregate concrete patio.

A pry bar was used to raise the door frame, while a wood shim was put in place to provide accurate level.

With the frame secured in place, doors were inserted and finish hardware was installed.

Special nailing clamps supplied with the door simplified securing the unit to the wood surround.

Interior Doors

Interior doors used for the three additions included standard swing units, bifolds for the garage conversion storage wall, and a pocket unit in the bath area of the bedroom-bathroom addition.

Pocket Doors Most pocket door assemblies are prefabricated units consisting of header, pocket framework, face jamb, and hanging hardware. The door is purchased separately, in the desired style and width, with height usually based on a 6-foot, 8-inch door.

Pocket door frames include the header, shown here being nailed to the pocket frame, and the facing jamb.

With the header attached, the assembly is positioned in the rough door opening and nailed to studs at the closed end.

A plumb bob is used to assure straightness of the frame, and a level is used to make certain the header is level before it is nailed in place. Wood blocks holding the frame apart remain until placement of the door in the unit.

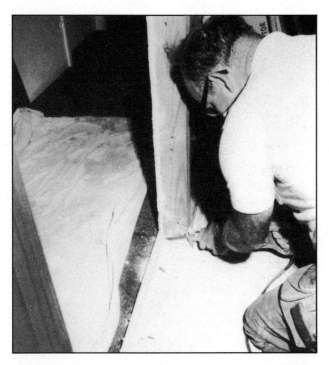

The 2 x 4 plate of the original wall was cut with a saber saw and removed in constructing the new passageway door. Finish framing included a prehung door, built to standard size in a local shop.

Swinging Doors In framing for regular swinging doors, remodelers can select assembled prehung units or prefabricated units delivered to the job-site knocked-down for fast assembly. The header is nailed to jambs at floor level, lifted into the rough opening, and shimmed with wood shingles until both jambs are the proper opening width and plumb. Most door frames are installed with shims placed at four or five locations along the jambs, starting from the bottom.

The original exterior front wall of the home was opened to provide a passage door to the bedroom hallway. Exterior sheathing was cut with a saber saw, and new framing was installed for the doorway, including header and support studs. The power saw then was used to cut the interior plaster wall neatly at the edge of the new framing.

After measuring the length of the header, a board guide was cut to this length and used during insertion of shingle shims placed at four locations along the jambs.

An archway passage between the bedroom and the bathroom areas was framed like a typical doorway, minus the door. Header and jambs were nailed together on the floor and set in the opening.

A plumb line was used to determine desired placement.

The Garage Door

The garage door switch from bypassing units to a full-width overhead door required a full workday for a single workman, plus further alteration when the automatic door controls were added. The job was complicated by a low headroom area not often encountered in garage construction.

Hangers were unbolted, and the old doors, overhead track, and lower slide guides were removed to begin the updating. A power saw was then used to cut the existing door jambs to the same width as the adjoining interior wall. This enables a flush fit of the new door.

Sectional doors such as the one selected here may consist of four- or five-panel units; the latter is recommended for durability. If a window or decorative glass is to be used, it will be located in the fourth panel from the bottom. The locking device is placed in the third panel.

One of the reasons for selecting a flush or paneled sectional garage door without glass is the added security it provides. Burglars have been known to break glass panels, reach inside, and unfasten the locking device, all in a matter of seconds.

Sectional garage doors are delivered to the site in sections for application of hardware and assembly at the door opening. Header and jamb beauty strips (stops) are applied flush to the inside surround, and then door sections are put in place.

The bottom door section is set in the opening and checked for level. If the floor is not level, as was the case in this installation, the panel is scribed and cut to fit. Succeeding panels are equipped with required struts and hardware and temporarily nailed in place at the opening. Nails are driven slightly into the jamb on each side and bent over the panel to hold it in place while working on other sections.

With all panels hinged together and temporarily secured to the opening, rollers are placed in the hinge openings, and the vertical track is put in place and secured to the wall or jamb. The horizontal track is then joined to the vertical track and secured to the ceiling. Placement of the torsion bar varies according to headroom.

Finish Flooring

Vinyl Sheet Flooring

Resilient flooring selected for bathrooms in two of the three remodeling projects was installed by traditional methods. The vinyl sheet material was obtained in standard 6-foot rolls and applied over particleboard underlayment.

Like all well-installed finish flooring, the job begins with a clean, smooth surface that will not telegraph cracks or nailheads. Seams in the new material should be kept to a minimum, and all necessary cuts made with a sharp knife.

Underlayment Standard ¼- and ⅜-inch-thick particleboard underlayment comes in a number of panel sizes, but most flooring applicators find the 3 x 4-foot dimensions the easiest to handle in bathroom work because of limited working space.

Particleboard underlayment can be applied with a staple gun, driven by mallet, or with 1¼-inch ring-shank nails spaced 4 inches on center. It is good practice to cut and fit all pieces before nailing.

Flooring applicators suggest that particleboard joints be spaced $\frac{1}{32}$ inch apart (or a knife blade thickness) to allow for possible expansion. They also advise that joints should be staggered. Spacing at the wall area is not as critical as other areas, inasmuch as this joint will be covered with wood cove.

While the underlayment is being applied, it is good practice to unroll the new vinyl material and lay it flat in an adjoining room, finish-side up. This will make the material easier to work with when cutting it to shape and size.

With the underlayment in place, the cove wood or "wood stick" is cut and mitered to fit the perimeter of the room where the vinyl will be extended up the wall the customary 4 inches. Again, all pieces should be cut and fitted before nailing in place with standard 1-inch lathing nails.

Silver- or gold-anodized cap metal is then applied at the desired height of the vinyl cove. This material is cut in the same manner as the wood cove and comes prepunched for easy nailing. The same type of cap metal is available for coving carpeting.

Snap a chalk line in placing the cap metal at the

Resilient Floorings

Type	Durability	Ease of Maintenance	Resilience	In-Room Quietness	Price
SHEET GOODS—Come in continuous rolls					
Vinyls Can be colorful, good resistance to wear, grease, and alkali. Some have asbestos-fiber backing; some have cushion backing.	Excellent to Good	Very Good to Good	Excellent to Fair	Excellent to Acceptable	High to Low
No-Wax Vinyls Special surface that makes upkeep easy, should not be waxed.	Very Good	Superior	Fair	Fair	High to Medium
No-Wax Roto-Vinyls with cushioned backing Same easy-care surface as above, plus resilient backing.	Very Good	Superior	Good	Good	Medium
Cushioned Vinyls Vinyl wear layer resists stains, cushioning increases resilience.	Very Good	Excellent	Good	Good	Medium
Cushioned Roto-Vinyls Roto-printed with vinyl wear layer on top. Some have vinyl foam between surface and backing.	Very Good	Excellent	Good	Good	Medium
TILES—Usually 12″ Squares					
Vinyl-Asbestos Vinyl resins with asbestos fibers.	Good to Superior	Variable	Fair	Acceptable	Low
Vinyl Composition No asbestos	Superior	Excellent to Superior	Fair	Acceptable	Medium to Low

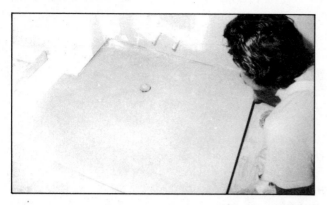

Vinyl flooring for the bathroom of the master bedroom-bathroom was installed by a flooring contractor. One-quarter-inch-thick, 3 x 4-foot underlayment was nailed to the subfloor, with the panels first prepared with necessary cutouts for plumbing requirements.

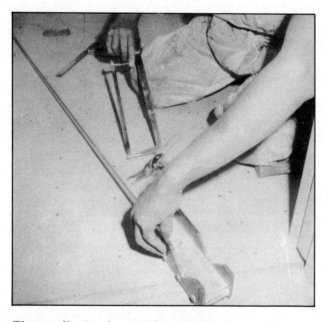

The small miter box used to cut the wood cove also was used to cut the metal cap which was installed 4 inches from the floor. A hinge pin was used to finish drive nails into the metal so as not to damage the lip that receives the new vinyl flooring.

A nail set is used to make certain that all nails are below the top surface of the particleboard underlayment.

correct height. This will ensure uniformity around the room, a factor of great importance if you choose a wall covering with a geometric pattern.

A soft bristle hand broom is very handy for cleaning the floor following the application of underlayment, wood cove, and metal cap. Rubbing the flat side of a hammer head across the underlayment should reveal any heads that require setting to obtain a smooth surface.

Three-foot-wide rolls of 15 lb. lining felt are used to prepare a room template for cutting the vinyl. The felt is trimmed to dimension, and cutouts are made for plumbing lines. Additional widths of felt are joined together and secured with tape to prevent movement while fitting and during material cutting. Making cross-marks in two sheets with a knife blade further assures accuracy, should the tape loosen at any point.

With the template prepared, reroll the new vinyl, this time with the face-side inward, and then unroll it again. This helps make the material perfectly flat for later cutting.

With all materials out of the new area, a patching compound, in this instance Vi-tex, is mixed on a piece of felt with the required amount of water. The compound then is used to cover seams, nailheads, and joints of the particleboard and wood cove. While the patching material is drying, the vinyl may be cut to shape and size.

A 6 x 9-foot roll of GAF vinyl selected for this bathroom blends well with the bedroom carpeting. The vinyl was trimmed on all edges, as factory-cut lines are not always true. A sharp flooring knife and circular cutters were used.

Multipurpose waterproof cement secured the vinyl to the particleboard underlayment. Once in place, a hand roller was used to eliminate bubbles, and a sealer was applied along the face of the built-in shower receptor, where wood cove was not used.

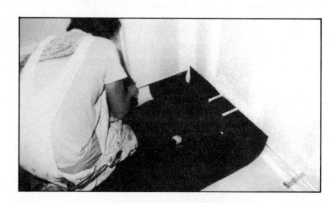

Standard 15 lb. lining felt in 3-foot-wide rolls is used to make a template of the floor surface, including cutouts for plumbing, cove angles, and wall corners. Adjoining widths of the felt are taped together to prevent movement, which would cause an error in cutting the vinyl.

After the felt template has been cut and fitted, a water-base patching compound is used to fill seams and cover nailheads. The material is mixed on a piece of felt, so the entire residue may be placed in a waste container.

A straightedge and cutting compass are used with the felt template to cut the new vinyl exactly to size and shape. The material is worked face-side up.

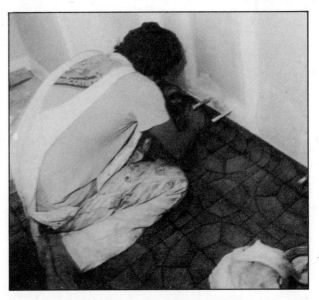

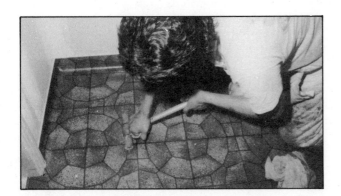

Pressure from a small hand roller secures the vinyl to underlayment and eliminates possible air bubbles. Note how the vinyl is extended to the center of the door jamb, where it will be met by carpeting. When the door is closed, only one floor material is visible from either side.

When the seam-smoothing patching material applied to the underlayment is dry and the vinyl is cut, the vinyl is applied to the underlayment with a multipurpose waterproof cement, working from the inside wall toward the room entrance.

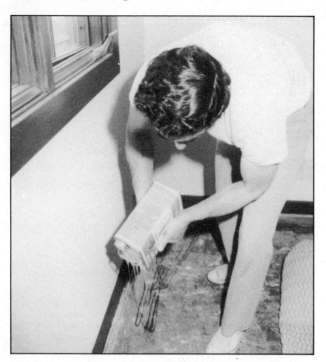

Following the nailing of carpet gripper strips along perimeter walls, a multibead trail of adhesive was poured to bond the rubber padding to the concrete floor.

The padding was cut and fitted within the carpet gripper strips, and all seams were covered with 2-inch-wide joint tape.

This view shows the wood carpet gripper strips and metal carpet cap after installation in the kitchen archway.

Closets were figured into the total dimension of the carpeting, thus eliminating seams. The turned-over sections shown here were fitted into the closet as part of the one-piece carpet in the room itself.

Carpeting

A single-pattern nylon carpeting was used for all existing and new floor surfaces throughout the major home addition, with the exception of vinyl flooring in the kitchen. Earthtone in color, approximately 197 yards of Evans Black Val d'Hor material was installed during the course of one workday by two professional carpet layers.

All floor surfaces were swept clean before the application of 4-foot-long carpet gripper strips along all perimeter walls of each room. Double strips were installed around the circular fireplace and at other areas where the carpeting runs more than 12 feet from the opposite wall.

The carpet gripper strips came with cement nails prestarted every 6 inches along the length to speed application to the concrete floor surfaces. All strips were positioned ¼ to ½ inch from the wall baseboard to permit "tucking" of carpeting during application.

Carpet layers suggest selection of a carpeting with double-jute backing because of its greater holding ability when secured by carpet gripper strips. In this installation, no additional tacking was used beyond the strips.

Following the nailing of the gripper strips, the

Required lengths were cut from large carpet rolls. Careful measurement was made along each side, and a small cut was made with a knife. A chalk line then was stretched between the cuts and snapped, leaving a line to be followed in cutting across the material.

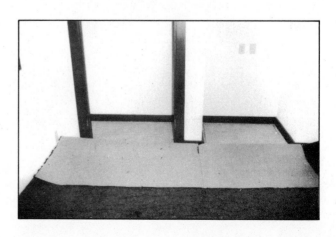

64-ounce rubber padding was fitted to each room layout. The padding comes in 4-foot-6-inch-wide by 75-foot-long rolls and is installed waffle-side-down with the fabric surface up.

All necessary padding seams (or accidental rips) are taped with standard paper pad tape available from flooring dealers and building-supply firms. The masking-tape-like material is 2 inches wide and is applied on the top side of the matting, which is fitted within the perimeter of the gripper strips and not over the strips.

Carpeting used for the major remodeling came in 12-foot-wide rolls and was cut to various room lengths on the concrete driveway, folded, and carried into the house. This cutting was done while the

rubber matting was setting in the special pad adhesive poured in multiple lines around room perimeters. Padding installed over wood subfloors on the second level was stapled in place, eliminating the need for pad adhesive.

A standard knee kicker and a power stretcher were used to install the carpeting throughout the house. Material was cut to larger-than-room dimensions and trimmed around the baseboard when fully secured to gripper strips. Seams were made with heat bond tape where required.

In carpeting the steps, double gripper strips were nailed 2 inches from the lower edge of the riser and 1 inch from the back edge of the tread. Padding was doubled in thickness to provide a greater fullness to the carpet installation and provide longer carpet life.

In all instances, the carpeting was laid so you look into the pile as you enter the room.

A power stretcher with adjustable pole is used in both directions in laying carpeting. The pole end (out of view) is against the opposite wall, as the workman stretches the carpeting by pushing down on the handle.

Heat bonding tape was used to join the carpet at seams. Supplied in rolls, this material is placed equally under each piece and heated with a small iron to create a long-lasting bond.

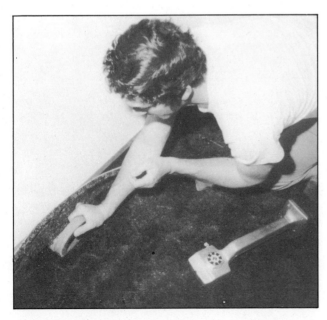

Once the carpeting is in place over the carpet gripper strips, a carpet trimmer removes excess material from the unfinished edge. The push-type cutter has a smooth edge to prevent damage to baseboard and walls.

A knee kicker is used to stretch carpeting a short distance and to secure a tight fit.

A sharp razor knife is used to trim carpeting from the jute side in fitting it around doorways and, as shown here, around the circular fireplace.

Stairs fitted with carpet gripper strips on risers and treads are shown here, ready for the padding.

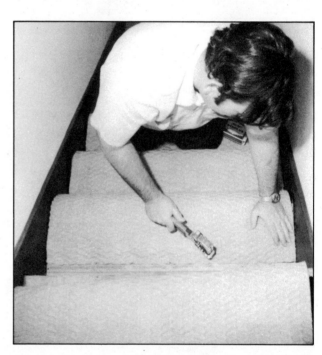

Double-thickness padding is stapled to the treads and risers, leaving the carpet gripper strips open to receive and secure the carpeting.

Bathroom and Sauna

Bathroom Plumbing

Each home addition in this book involved construction of a full bathroom. The typical residential plumbing system consists of these basic components:

- Soil stack—a 3- or 4-inch pipe which runs vertically from the lowest point in the system to 6 or more inches above the roof line where it is flashed for protection from the weather. The stack may extend more than one story, and when equipped with aerator fittings (for self-venting), a single stack can be used for waste and vent plumbing in multistory buildings such as apartments. Two stacks can be joined together at the top above the highest fixture with only one stack extending through the roof line.
- Drain pipes—usually 1½ inches minimum to accommodate bathtubs, shower stalls, lavatories, and laundry trays. Also called "branches," these pipes are joined into the vertical soil stack. While lavatories can be adequately drained with 1½-inch branches, water closets require a minimum soil branch size of 3 inches, and both bathtubs and shower stalls are best served by 2-inch waste branches. It might also be noted that water closets should be located with a minimum of a 15-inch distance from the center of the fixture to the wall or cabinets at each side. A minimum space of 24 inches should be allowed from the front rim of the water closet to the facing wall surface.
- Vent stacks—providing for a flow of air to or from a drainage system or forcing a circulation of air within the system to protect trap seals from siphonage and back pressure. Vent stacks are usually 2 inches in size and run from the horizontal drain line at the base to an intersection in the vertical soil stack. This latter connection must be at least 6 inches above the overflow rim of the uppermost fixture served. Vent stacks continue through the roof line and may be no closer than 10 feet to a window, door, opening, air intake, or ventilating shaft. The vents must be 10 feet above the ground level and 6 inches above the roof line. The vent must also be at least 3 feet in every direction from any lot line. Fixtures without vents can, under certain conditions, siphon water from the traps

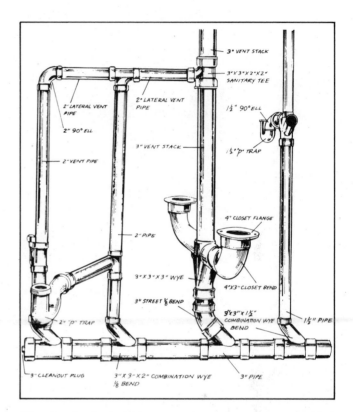

The plastic pipe DWV system shown here was assembled with all PVC pipe and fittings at a savings of approximately 70 percent when compared with metal pipe and fittings. Note the 3-inch vent stack which permits standard 2 x 4 stud wall construction. Larger 4-inch stacks require greater wall thickness.

and allow sewer gases to enter the room.

- Supply lines—typically ½-inch copper pipe with the hot water line installed vertically a minimum of 6 inches to the left of the cold water line when the faucet side is viewed from the front. Lines from the outdoor meter to the water heater are generally ¾ to 1 inch in diameter.
- Traps—curved pipe or tubing shaped like the letters S or P which, when properly vented, produce a liquid seal which will prevent the back passage of air without materially affecting the flow of sewage or waste water through it. Approved connections are measured from the trap to vent along a horizontal line, except a water closet or bidet, whose measurements include the developed distance from the top of floor flange to the inner edge of the vent.

The garage conversion locates the bath back-to-back with the family bathroom in the existing structure. In this installation, a standard-size tub was installed along with a water closet, and a single lavatory bowl was built into a new ceramic vanity top. Water supply was achieved by tapping into existing lines, and new waste lines were connected to those located in the family bathroom.

The new major addition second-story bath locates the water closet and tiled shower opposite one another in a separate compartment entered from the dressing room, which includes a twin-bowl vanity arrangement. Supply and waste lines were connected to those located immediately below in the laundry area of the attached garage. The tiled shower stall was fabricated at the site with workmen using three sheets of 15¢ building paper and three coats of tar (hot mop) to form the shower receptor, which in turn was tiled along with the water-resistant gypsum wallboard walls and the ceiling of the stall. Tile also was used to surface the built-in lavatory countertop.

The ample-size bath area of the bedroom-bathroom home addition includes water closet, bidet, and 5′ x 2′10″ dual-bathing shower, fabricated with a precast fiberglass receptor and Corian walls. The twin-bowl built-in vanity with Corian top is located in the dressing area, opposite the walk-in wardrobe closet and around a partition wall from the sauna.

Acrylic panels, used for the shower stall walls, required a minimum of on-site cutting to fit the area, plus drilling for the shower head and single-handle water controls. The panels were laminated to water-resistant wallboard with adhesive and caulked at the joints.

The dual-bowl vanity top came prefabricated and was set atop a locally made wood vanity. Mastic was used to secure the acrylic unit to the vanity top support and also to attach the backsplash to the countertop surface.

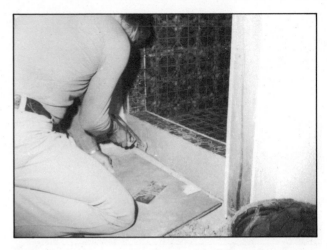

Ceramic tile, in 6-inch squares, was used for the interior of the shower stall, while 4 x 4 units were applied to the inner door frame.

Shower stall panels were cut to the desired length with a power circular saw. The plywood panel clamped to the acrylic panel served as a cutting guide.

The hot mop shower receptor constructed for the second-story bathroom is shown here filled with water, ready for plumbing inspection. After approval, the base was finished with a combination of mortar, wire reinforcement, and tile. The floor slopes slightly from all four sides to the center drain.

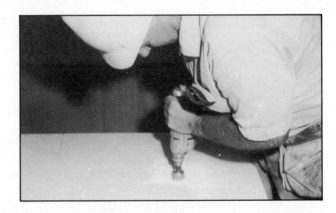

A power drill equipped with circular cutter was used to provide a hole for the shower head arm extension.

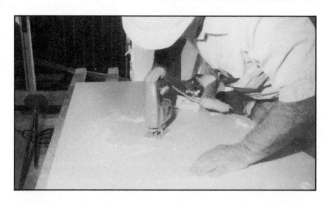

After drilling a small hole at the edge of the shower control location, a jigsaw was used to remove the circular area.

Shower stall panels were set atop the ledge of the precast shower receptor and were braced with wood until the bonding adhesive has completely set. The cardboard at the base serves as a protective cover for the fiberglass receptor.

Adhesive was applied both to gypsum wallboard and to the plywood used behind the shower stall panels.

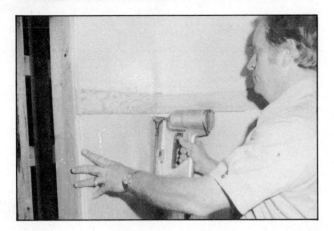

With 1 x 3 wood furring strips nailed to the studs and existing walls of the new sauna, the workman used a power staple gun to apply styrofoam insulation in the voids.

The walls and ceiling then were covered with 36-inch-wide sheets of aluminum vapor barrier, stapled to the furring strips.

With the room totally wrapped in vapor barrier (except for the floor), the workman applied tongue-and-groove ceiling panels, blind nailing each board along the tongue.

Sauna

Installation of a sauna can be a one-day job for an experienced handyman using a prefabricated unit. The Viking sauna selected for the bedroom-bathroom addition was constructed in a 6′4″ x 9′ area adjacent to the bedroom dressing area and new bathroom.

Standard 2 x 4 stud partition walls were used on two sides, the original house wall (stripped to sheathing) on one side and the new exterior wall of the home addition on the other. Inside this space, the packaged sauna was assembled, piece by piece, as cut and packaged at the factory.

The installer first applied 1 x 3 furring strips horizontally across the inner walls and studs, leaving 2 feet between rows. Styrofoam insulation panels 1 inch in thickness and 2 x 8 feet in dimension then were power-stapled to the framework between nailing strips.

Still using the stapler, the workman covered the entire room from floor to ceiling with two 36-inch-wide sheets of aluminum vapor barrier, lapping it at the joints.

Kiln-dried redwood tongue-and-groove panels were applied to the ceiling of the sauna, using 1¼-inch staples for blind nailing. The walls were surfaced in a similar manner, and a sheet of ¾-inch particleboard was placed over the floor area in front of slanting framework for the built-in bench. Additional redwood tongue-and-groove panels were stapled to the bench framework, and the benches of 1 x 3 redwood fixed in place.

The sauna heater and controls were mounted directly across from the two-level benches and required a 120V, 20 amp supply line. The preassembled door required a rough opening 80½ inches high and 26 inches wide.

Vertical redwood boards were applied in the same method as ceiling boards, and 2 x 4 framework was installed for the long-dimension benches. A sheet of particleboard then was applied in front of the benches, later to be surfaced with indoor/outdoor carpeting.

With the carpeting in place, additional tongue-and-groove redwood boards were blind nailed to the slanting 2 x 4 framework and fully nailed to the wall paneling.

A power stapler again was used to secure the prefabricated benches to support members.

The precut sauna door framework was quickly assembled and set in the rough opening. Shingle shims were used to obtain the proper fit, and with header and jambs nailed securely in place, decorative wood trim was applied.

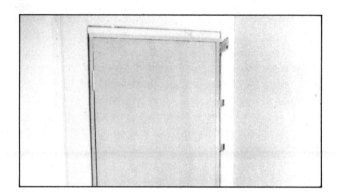

The sauna heater and controls are located opposite the long benches; they required a separate 120 line from the new master electrical control panel installed in the adjoining closet. Temperature and timing controls are located above the heater, which is fenced in for safety reasons.

Cabinetry and Millwork

Cabinet Selection

With the exception of shelving and the cabinet and drawer units custom-built into the storage wall of the garage conversion, all other cabinetry involved was factory-made, either locally or out-of-state by modular-cabinet manufacturers.

Master Bedroom-Bathroom The bedroom-bathroom addition required a single vanity unit in the dressing room. Purchased from a local cabinet shop, the unfinished birch unit was constructed for dual-bowl lavatory installation, with a door and storage area below each bowl and a center row of drawers from countertop to toe space.

In selecting a unit of this type, it is important that the manufacturer be advised of your choice of a finishing method. In this instance, a mixup occurred, and several drawers and doors had to be replaced to gain the uniformity required for staining the wood. Had the owners intended to paint the face of the unit, the gradation in wood grains would not have been so important.

The 6-foot, 1-inch-long vanity was delivered to the job site ready for exact placement between bath and bedroom walls. The subfloor was level, so no further adjustment was required. The vanity was nailed to studs of the rear wall and then topped with the prefabricated countertop-bowl assembly.

Major Home Remodeling The major home expansion required all new cabinetry for the remodeled kitchen and a combination vanity-wardrobe unit for the new master bath-dressing area.

Prefinished, modular cabinetry was selected for the kitchen and put in place following the taping of the gypsum wallboard, before the installation of vinyl flooring.

As is the situation in most remodeling and home expansion projects, the owners shifted their thinking somewhat during construction, and this caused a change in the original plans for the kitchen. Original plans to keep the existing built-in cooking units were scrapped as the owners more fully envisioned their new kitchen layout and its proximity to the family room addition and the living-dining room.

Instead of having a separate oven and cooktop, the owners chose a new combination slide-in unit and relocated the refrigerator-freezer to the original oven location. This resulted in a minor change in cabinetry selection. At the same time, a planned built-in opposite the fireplace in the family room was eliminated to provide better foot traffic. The pass-through area from kitchen to family room was redesigned to provide better visibility, requiring a change in cabinet sizes. Still another change was the elimination of a planned sliding pocket door between the kitchen and dining area.

The final layout for the kitchen called for standard modular base cabinets flanking the slide-in oven-range and an L-shaped base cabinet arrangement for the sink work center and serving bar.

Modular base units were installed first, with care taken to make certain all units were level and at a perfect 90° angle where they meet to form the "L". Wood shingle shims were used to level units on the concrete floor and to shim the units away from the

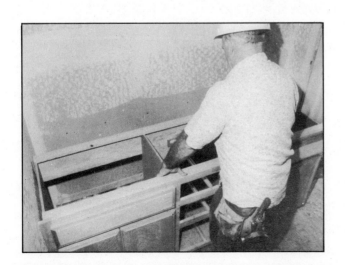

The twin-bowl vanity for the bedroom-bathroom addition was set in place between the bathroom and bedroom walls. The unit was nailed through the back framework to stud framing.

Kitchen base cabinets were set in place and leveled side-to-side and front-to-rear with wood shingle shims. A straight board was used to line up cabinets flanking openings for an automatic dishwasher and combination oven-range. Note also that wood shims were installed between cabinets to maintain proper alignment.

wall where a flush front was needed.

With all of the base units in position, side-by-side cabinets were attached to one another at three locations through the front stiles. "C" clamps were used to hold the units in place before and during nailing.

The same procedure was followed in attaching wall units to studs and to each other. Unfortunately, the finished installation required a further visit from the manufacturer's local distributor to correct improperly hung doors that did not match up edgewise across the bottom surface.

With all of the kitchen cabinets in place, a ⅝-inch plywood panel was applied to the top of the cabinets to serve as a base for the ceramic tile countertop.

The master bath vanity and wardrobe units were fabricated locally as individual units and were joined together at the site before being stained along with other wood trim in the house. The single-bowl vanity unit includes two drawers above door openings and is topped with a ceramic tile countertop.

The adjoining garment drawers (a total of eleven) are located from toe space to a 24 x 24-inch, one-shelf storage area at ceiling height and are closed off by a pair of doors. A single door conceals all of the 6-inch-high garment drawers.

Other shelving, such as in wardrobe closets and linen cabinets in this home addition as well as in the others, was cut to size at the site and installed. Closet rods were positioned at two heights to better utilize space. A framework was planned for the closet of the bedroom-bathroom addition to accommodate stereo controls, to be added later. This closet also was lined with cedar and contains the hatch opening to the expanded attic area.

"C" clamps were used to hold cabinets in correct position for nailing.

The completed cabinet installation shows the work center area with sink located to the left of the automatic dishwasher installed in the base opening. A lazy-susan cabinet was used in the corner area adjoining the serving bar at right. Additional wall cabinets were suspended from the soffit above the serving area and may be opened from either side. Cooking-refrigeration centers are out of view at left.

Kitchen Cabinet Basics

Kitchen cabinets are among the most efficient or inefficient storage elements of the home depending upon the type and number installed and their specific interior arrangement. You can choose from several woods, decorative plastic laminates, metal finishes, and decorative styles to achieve the specific look you desire and the proper amount of storage for your family.

Standard base and wall cabinets are made in varying widths in 3-inch modules from 9 to 48 inches. Those manufactured in a factory usually have sides and backs plus adjustable shelves. Custom cabinets can be made at the job site, but in most instances today you can obtain completely finished furniture-style cabinets for less money when they are factory made.

Stock-size cabinets range in price from lower to upper medium and often fill all remodeling needs. Custom units, made to specific kitchen measurements and requirements, cost more, yet afford the buyer a full selection of material, size, and type. Fancy moldings and special decorative finishes and interior fittings can increase the cabinet price.

Wood and plastic-laminate cabinets account for the largest share of the market today, with metal units a far-distant third. Manufacturers and kitchen specialists offer detailed literature on specific lines, and local cabinet manufacturers will be pleased to show you samples and explain construction and use features.

There are four basic kitchen cabinet styles: traditional, colonial, provincial, and contemporary. Each has a distinctive design that sets it apart and establishes the mood of the entire kitchen. For more information on cabinets and styles see *Kitchen Planning and Remodeling* and *Space-Saving Shelves and Built-ins*.

Traditional styling continues to be the most popular with American homeowners. These cabinets usually have a recessed or raised panel door and drawer styling.

Colonial or Early American styling can be knotty pine or more elaborate pegged and vee-grooved door facings. Distinctive hardware (hinges and pulls or knobs) further establish the period.

Provincial styling generally is French or Italian with delicate moldings on the face of the doors and drawer fronts. In some cabinet lines a routed groove is used instead of the molding.

Contemporary styling traditionally employs flush doors and clean lines. Often the doors have what is known as a reverse lip which eliminates the need for hardware pulls, and the hinges are concealed and not visible.

In addition, Mediterranean styling has become popular in recent years. Usually dark finished, the style is heavy and sometimes highly ornate.

Regardless of the styling, kitchen cabinets may be said to consist of five elements: (1) base units, (2) wall units, (3) drawers, (4) doors, and (5) miscellane-

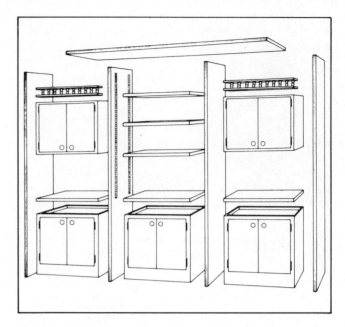

The term "kitchen cabinets" has become a misnomer as builders increasingly utilize them throughout the house. Most manufacturers offer components to adapt them. These adapting components and ways to use them are from a booklet from Rutt Custom Kitchens.

ous, such as bread boards, chopping blocks, and pull-out vegetable bins.

Base cabinets vary in depth from 22 to 24 inches, from the face of the cabinet to the wall, and are 34½ inches high without the countertop. Most countertops are 1½-inches thick, bringing the total base cabinet height to the recommended 36 inches. Dishwashers and most ranges are designed for the 36-inch counter height, although some range tops may be positioned 2 or 3 inches lower than its side rim. Some homeowners find it convenient to have a section of countertop area at a 33-inch height, usually at the end of a countertop run.

Base cabinets may be obtained with right or left-hand swinging doors or both; with drawers only; or with a combination of drawers, doors, and shelves. Specific units are manufactured for single and double-bowl sinks, for corners where a revolving shelf arrangement makes full use of otherwise unused space, for built-in ranges, and for special accessory use such as pull-out bins and shelves.

Wall cabinets are 12 to 13 inches deep and are usually hung with the top line 84 inches above the floor to match the 84-inch height of utility (broom) cabinets. The height of the basic wall cabinet may be 30 to 33 inches, depending upon style. This gives ample clearance (15 to 18 inches) between the counter (36 inches from the floor) and the bottom of the wall cabinet.

If a wall cabinet is to be placed over a sink, allow 24 inches above the sink rim for the standard-depth wall cabinets. This space can be reduced to 16 inches if the wall cabinets are custom-made only 6 inches deep. Local codes should be checked when installing cabinets above the range. Usually there is a clearance of 27 to 36 inches and depth should allow for the range hood. Specific wall units are made for use over freestanding refrigerators and freezers.

Oven cabinets customarily have three drawers below the appliance and a two-door storage area above the appliance. Widths vary from 24 to 33 inches, and in many floor-to-ceiling units, the height can be cut from 84 to 82 inches if necessary.

Manufactured kitchen cabinets can be purchased with a number of "trim" items useful in filling an exact lineal or vertical dimension. Face molding, outside corners, trim molding, scribe molding, valance panels, and end cabinet panels frequently are required to complete the installation. Likewise, quarter-round and half-round shelves are available for full-view cabinet ends.

Among the many cabinet accessories offered for kitchen use are a metal bread drawer, cutting boards, combination soap and dishcloth rack, pan rack, suspended cutlery trays, extension towel racks, flour bin, spice racks, sliding storage bins, extension pan holders, extension wastebasket, tray base, vegetable base, food mixer mechanism and shelf assembly, and full-depth drawer extensions.

Kitchen desks or planning centers are offered in matching cabinetry. Units provide a center drawer and come with or without drawers on either or both sides. The kitchen desk should be installed about 29 inches off the floor, out of the traffic pattern, and away from the spatter of the range or sink. Shelves can be installed above for cookbooks, and frequently the telephone is located here.

Kitchen cabinet doors are of two principal types—flush and panel. The flush doors may be hollow core or solid core, or simply a panel of plywood. If only a panel of plywood, it is generally ¾ inch thick. Some cabinet doors, either panel or flush, are rabbeted or lipped while others have a square edge. Self-closing hinges and magnetic catches are highly desirable.

Kitchen cabinet drawers generally vary in depth from 18 to 22 inches to satisfy the particular depth of the base cabinet. For greater ease of operation, drawers are often provided with metal slides and roller bearings or small wheels which run on metal tracks. Otherwise, drawers slide on wood drawer runners. Fronts are flush or lipped.

If kitchen space permits, the old-style walk-in pantry can be highly useful for the storage of canned and packaged foods, small appliances, dishes, and glassware. When wall space is at a premium, use pull-out shelves that bring all the stored items within easy reach. Or you can hinge shelf

sections on both sides so they will open like the pages in a book to make everything accessible. Slanting shelves, supermarket style, can also be used for canned goods storage with cans lying on sides so that a replacement rolls into position as one is removed.

Millwork

Newly paneled rooms can be given the professional look by completion with the appropriate moldings—for finishing door and window openings, at inside and outside corners, where paneling meets the ceiling, as a cap to cover exposed panel edges (such as the top of wainscot panels) and at the floor line to cover the gap between the bottom edge of the paneling and an uneven floor.

Standard molding patterns are available in various species of wood including ponderosa and sugar pine, white fir, douglas fir, and larch. Pine moldings are also available in many areas in specified length finger-jointed moldings for use in applications where the surface will be covered with an opaque finish.

In addition to unfinished wood moldings sold by most lumber and building material dealers, prefinished moldings are offered in colors and wood grains to match or harmonize with various wood panels. Some of these moldings have a vinyl outer skin that never needs refinishing while still others are embossed cellular vinyl with the color going all the way through the material.

Standard moldings have particular applications in finishing a room as shown here. Specific shapes are made prefinished and unpainted for job-site finishing. Other molding patterns also may be obtained from some companies on special order.

Millwork and trim used in the garage conversion and the bedroom-bathroom home addition were purchased locally from retail lumber dealers. The millwork for the major addition for the most part came from a similar source, but the rest was ordered from an out-of-state manufacturer, particularly the wood French doors.

Standard baseboard moldings and door trim were applied in the two smaller remodeling projects. The third job required 1 x 4 wood baseboard in every room and a 1 x 6 wood ceiling trim throughout the house. Door trim also was 1 x 6 wood on both sides of the jambs.

Turned spindles of stock design and dimension were used to provide a 3-foot-high railing inside the front entrance, separating the new tile walkway from the living room. The spindles were joined to a 3½-inch-wide top and base rail.

Original plans for placement of 4 x 4 wood beams were changed somewhat in the actual installation,

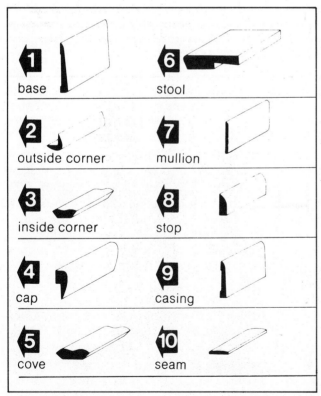

Standard moldings have particular applications in finishing a room as shown here. Specific shapes are made prefinished and unpainted for job-site finishing. Other molding patterns also may be obtained from some companies on special order.

with a couple runs being eliminated and the fastening method completely revised. Rather than going through the face of the beam with hex-head lag screws, workmen secured the beams in place from the attic side so no fastener would be visible. Each beam was secured to concealed framing with wood screws. The beams then were trimmed with ¾ x 1½ stock nailed to the beams at the ceiling edge. This same trim was added to ceiling moldings in other rooms of the home.

Painting and Finishing

In the garage conversion and the bedroom-bathroom, the homeowners undertook all painting and finishing tasks.

The Philippine mahogany paneling selected for the garage conversion was purchased unfinished and then stained. After sanding with fine sandpaper, the coating was applied with a brush and wiped with a soft cloth. A final coat of flat varnish completed the surfacing. Doors of the storage wall were also finished in this manner.

The exterior redwood siding was coated with Behr Plus Ten clear water-base finish, matching the panels with those of the existing exterior siding.

Water-base stucco paint was used for the exterior of the new bedroom-bathroom home addition, and latex paint also used for interior walls. Contrasting paint was used for wood trim throughout the addition. Walls and ceilings were "cut" with a brush and then painted with a standard roller.

Professional painters were employed to finish the major home addition, applying coatings according to the general specifications.

Vinyl-coated, cloth-backed wall covering was selected for redoing the walls of the existing family bathroom in the major home expansion. The material was applied after new woodwork was in place and finished. The water closet was removed during the wall covering application.

Paint applied to the exterior surfaces was brushed on all trim and in areas not large enough for roller application. This area is the front of the major expansion, where new windows were added. The porch overhang was surfaced with stucco.

Paint rollers were used to apply coatings to the new and old stucco.

The new second-story master suite was painted primarily from ground level by means of a roller and extension handle. A drop cloth was used to protect the shake overhang above the garage entrance.

Wood stain was brush-applied for all interior trim, sash, and doors throughout the major home remodeling.

Specialties

Fireplaces

The major remodeling project involved both the refacing of an existing fireplace in the living room and the installation of a prefabricated circular fireplace in the new family room. (Note: For installation and selection information on all types of fireplaces, see *Homeowners Guide to Fireplaces*.)

The refacing operation required removal of a tile surround from the brick substructure and installation of new stone facing, raised hearth, and mantel. The existing hearth was used as a base for the new stonework, which nearly doubled the width of the original fireplace setting.

With the facing tile removed, the stone mason secured metal lath (similar to that shown for stucco walls) to the brick understructure, fastening it with masonry nails. Corrugated metal ties then were nailed to the brick and adjoining gypsum wallboard. Spaced no more than 2 feet apart, these ties were used to help hold the new stone and grout in position

Working upward from the hearth, the workman arranged three color varieties of stone in pleasing fashion, grouting the units with a mixture composed of three parts No. 1 or No. 2 sand, one part regular cement, and one-half part fire clay.

Stonework for the hearth was completed in similar fashion, with the design calling for a higher elevation on the left side than the right side of the fireplace opening. Some of the vertical stonework required face-bracing with wood until the mortar set.

The interior of the fireplace, as well as existing vent and chimney, required no alteration. The mantel was added along with finish wood trim applied throughout the house.

The new prefabricated circular fireplace constructed in the family room actually began with the forming prepared for the concrete slab foundation, when a recessed area and gas-line raceway were poured.

Inside this circle, workmen formed a raised stone hearth using the same stone varieties selected for the living room fireplace refacing. Cinder blocks were arranged at the base to correspond to the octagonal shape of the factory-made metal fireplace. Stones were set atop the blocks, again using the same mortar mix as employed for the living room unit.

The stone surround was allowed to cure for several days before the interior base was covered with sand and a layer of yellow firebrick, laid to a height one-brick short of the top of the surrounding stonework. Placement of this brickwork correlated exactly to a metal template prepared to the exact measurements of the fireplace base.

Once the entire base had cured (three to seven days, depending upon humidity), the prefabricated 42-inch-diameter fireplace was set in place by the local manufacturer. The black-and-brass unit required an octagonal cutout in the new gypsum wallboard ceiling for installation of the chimney.

UL-approved 10-inch ID interlocking metal chimney sections were attached to the fireplace and extended to a height 2 feet above the roof line. It might be noted that a 3-foot height would have been required if the vent were less than 10 feet from the ridge line of the house.

The flue assembly consisted of two 36-inch-, one 24-inch-, and one 18-inch-long sections plus a terminal cap with spark arrestor. This assembled unit was kept a minimum of 2 inches from all surrounding combustible materials.

On the exterior assembly, the flue was located in a 3-foot-square chimney housing framed with wood, flashed at top and base, and stuccoed.

The finished fireplace was then completely wrapped with lightweight film to protect exposed surfaces during painting.

Smoke-stained tile facing of the living room fireplace was removed and replaced by attractive stone. The existing structure of the fireplace required no further alteration.

Three varieties of stone were used to create the new surround and two-level hearth.

Wet mortar joints are brushed smooth and free of excess material following careful placement of the firebrick base that corresponds to the prefabricated fireplace.

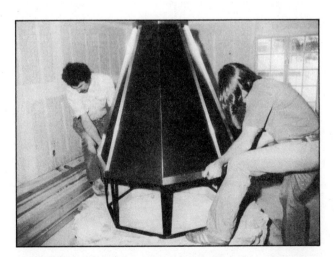

The prefabricated fireplace is shown here being set atop the firebrick platform, to which it was then secured with masonry screws. Snap-together sections of the triple-wall UL-approved chimney were further secured with sheet-metal screws.

A reinforcing rod is positioned inside the stone hearth of the fireplace and becomes an integral part of the firebrick base.

Cinder block and attractive stone were used to construct the raised hearth of the new family room fireplace. The string line designates the center of the 42-inch diameter unit. Piping is for the gas lighter, and the wood block is employed as a brace while the mortar sets.

An octagonal cutout was made in the family room ceiling to accommodate the prefabricated fireplace stack and adjoining flue.

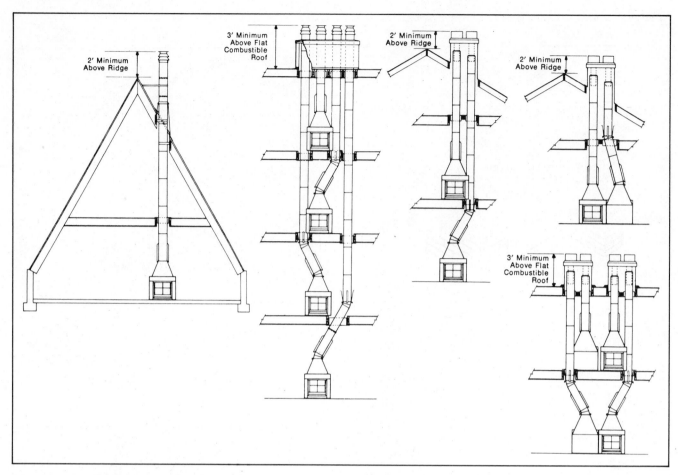

Various applications of prebuilt fireplaces. Drawings courtesy of Heatilator Fireplaces

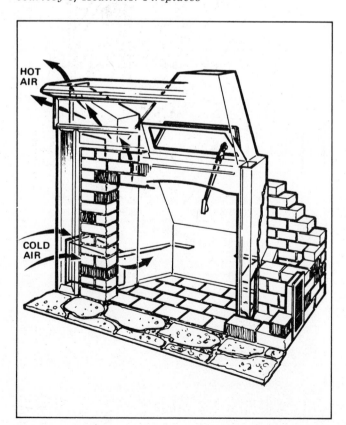

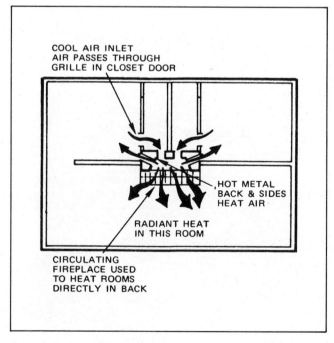

This drawing shows a typical circulating fireplace. The cold air is drawn in the bottom, heated between the two steel walls, and moves out the top.

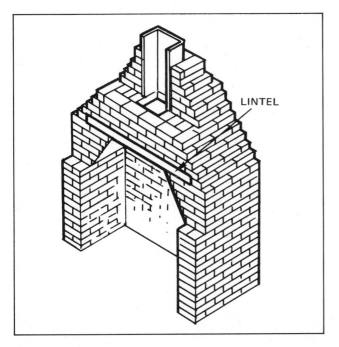

Shown here is the method of building a fireplace where the firebox is built after chimney construction. Note the lintel over the smoke chamber and the support for a breast lintel to be installed later. With this method of construction, the entire fireplace front may be torn out at some future date for remodeling without endangering the structure. Note also the support for the flue lining.

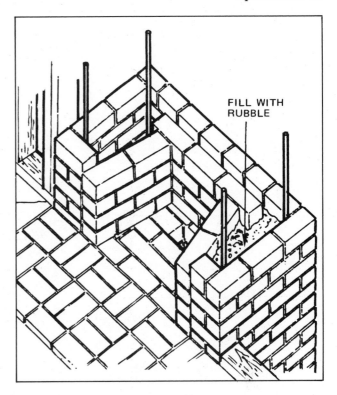

Reinforcing bars at firebox level

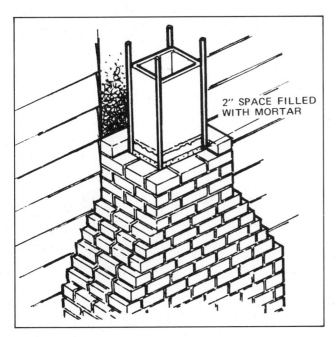

Reinforcing bars at flue level

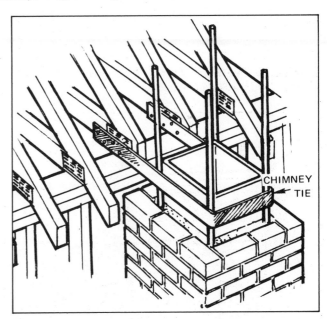

Anchor ties at roof line

The new family-room chimney housing is 3-foot-square, with the metal flue held more than 2 inches away from combustible materials. The housing has a metal terminal cap with spark arrestor.

Space for the new stairway to the second-story master bedroom was obtained by reducing the size of the downstairs bedroom, which immediately adjoins the garage. New wall framing was constructed and surfaced with fire-resistant gypsum wallboard.

Shown here is the opposite wall of the new wall partition, built to enclose the stairway. Blocking between studs was used to receive nails driven through stair stringers on the opposite side of the wall.

Stairways

An open stairway is a good choice if you want a spillover of light from the upper floor to add to the general illumination of the lower level. Half-walls along the length of the staircase provide protection for an open staircase. An enclosed staircase is the best choice if you have no wall space.

Regardless of the type you choose, be sure you have four-feet clearance at the top and bottom, and light switches at the top and bottom of the steps. Proper railings and treads are essential safety factors.

The major home remodeling added a stairway to the new second floor, using the area under the stairs for storage.

Rough framing for the stairway is shown here partially completed. The new area under the stairs is now used for storage, reached through a crawl door located in the den wall.

Glossary

Acoustical Material's ability to lower noise transmission or absorb sound.

Acrylic Resin A base material used in some latex paints and caulkings.

Anchor Bolts Embedded in masonry to secure framing.

Asphalt Waterproof vehicle derived from petroleum and used in roofing materials.

Awning Window A projecting window, hinged at the top, opening up and out like an awning.

Backfilling Moving excavated earth back into space adjacent to foundation walls.

Baseboard Molding that covers the joint between floor and wall.

Battens Narrow wood strips to cover joints.

Batter Boards Horizontal boards paired at building corners for foundation grade and alignment.

Bay Window Composed of three or more individual windows, generally with the side or flanker units at 30° or 45° angles to the wall; a Bay projects from the wall of the structure.

Beam Load-carrying member that is end-supported and on which joists or rafters rest.

Bearing Wall A wall that carries the load of a roof structure.

Beveled Siding Horizontal strip material with bottom edge lapping top of the strip below.

Bifold Doors Doors hinged in two or more places for folding partially or completely out of the door opening.

Blanket Insulation Roll form of glass fiber or mineral wool material in widths to fit framing. Batt form is same material in short lengths.

Blocking Small wood pieces used to brace framing members, or to provide a base for sheet materials, or a fire-stop nailed part way up a wall, between studs.

Bridging Wood or metal crosspieces which keep joists braced and aligned in vertical position.

Cantilever Floor or structure extending outward beyond vertical supports.

Casement Window Hinged for opening on a vertical edge, like a door.

Casing Molding concealing joint between door and window edges and adjacent walls.

Caulking A flexible material used to close cracks or spaces and make them waterproof.

Ceramic Tile Flat fired clay units with glazed or other finish for surface durability.

Chalk Line Mason's cord saturated with fine chalk dust for use in making snap line marks.

Circuit Electric wiring from current source to point of use and return.

Circuit Breakers Switch-like devices that protect circuits from overload.

Cleanouts Waste or drain fittings with removable plugs for access to obstructions in the line.

Color Coding Electrical conductor identification to facilitate proper connections.

Concrete Hard-setting material made from Portland cement mixed with sand, gravel, and water.

Conductors In electrical circuits, the copper current-carrying wires.

Conduit A metal tubing system for carrying electrical circuits.

Corner Bead Metal molding to reinforce interior gypsum wallboard or exterior stucco wall corners.

Cornice Enclosing trim at juncture of roof's eave and rake.

Course Horizontal row of shingles, siding, or masonry units.

Crawl Space Space between floor and ground.

Cripple Framing members that are less than full length, as over or under a window.

Curing Concrete Keeping surface moist for initial period.

Diffuser Top part of a heating register having adjustable grille for directing air flow.

Distribution Panel Central circuit connection box containing protective circuit breakers.

Double-Hung Window Two vertical sliding sash which bypass each other in a single frame.

Downspout Drainpipe on exterior wall to carry roof water from gutters to ground.

Drywall Common term for gypsum wallboard, in contrast to wet-applied plaster.

Duct Rectangular or round air-carrying pipe in heating and air conditioning systems.

DWV Piping Plumbing lines for drainage, waste, and vents.

Easement Legal right of one party to make use of land owned by another.

Eave Lowest projection of a roof overhang.

Elevation Drawing that shows a straight-on view.

Excavating Digging into ground for foundations.

Exhaust Ducts Used with exhaust fans or ventilators.

Expansion Joints Bituminous strip material used with concrete slabs to permit slab expansion.

Fascia Roof edge facing board nailed to rafter tails.

Fiberboard Medium-density building board made from wood or vegetable fibers having some insulative properties.

Fixture Trim Faucets and other related devices that fit on plumbing fixtures.

Flanges Protruding pieces on beams, added for strength or used for attachment.

Flashing Metal or composition material that protects construction from or diverts moisture.

Floating Initial leveling of a concrete surface to bring it to a coarse finish.

Flue The space in a chimney through which fumes (smoke, gases) travel to the outdoors.

Footing Flat, thick concrete slab serving as base for foundation wall.

Foundation Wall Bearing wall of concrete, masonry, or wood for supporting first or ground floor.

Framing Structural wood skeleton and supplements.

Furring Strips Wood strips applied to walls or ceiling as nailing base and leveling method.

Gable End Triangular-shaped exterior wall immediately below edges of a gable roof. Frequently a space for application of an attic vent.

Gable Rafters Paired end rafters of a gable roof.

Grading Scraping and shifting the ground surface around a home's exterior.

Grout Thin mortar cement for filling spaces between ceramic tiles, or any thin sand-cement mix.

Ground Fault Interrupter Shock-protective device for wet-area or outdoor electrical circuits or outlets.

Grounding Connecting an electrical system to earth.

Gutters Metal or plastic troughs along roof edges for collecting roof water.

Gypsumboard Sheet material with gypsum core between two paper faces.

Hardboard Sheet, panel, or strip made of wood fibers compressed under heat to a dense board.

Hardwood From broadleaf trees such as ash, maple, or oak used in flooring, cabinets, and paneling.

Header Framing members that span an opening and support free ends of other members.

Hearth Front floor area of a fireplace.

Hip Framing To form hipped roof having same slope or pitch on all four sides.

Hollow-core Door Lightweight door with interior air space, for interior use.

Hub Bell end or a bell-and-spigot type clay or cast-iron sewer pipe; newer pipes that join by adhesive strips are called "no-hub" sewer pipes.

I-Beam Steel beam whose cross-section shape looks like an I because of its flanges.

Insulating Board Insulative sheets that serve as sheathing boards.

Insulation Light-density material that reduces heat transmission; also the protective covering over electrical conductors.

Jambs Members at the sides of a door or window opening.

Joists Horizontal framing members of floor or ceiling.

Kiln-Dried Lumber Wood factory or mill-dried to desired moisture content, minimizing warp and shrinkage.

Knee-Wall Low-height wall below rafters in attic.

Knockout Small, round metal part of an electric box, provides opening for cable or conduit connections.

Laminate Layered material made under pressure.

Ledger Horizontal board on which other framing members can rest.

Lintel Load-carrying member of steel or reinforced concrete used over masonry openings at doors, windows, fireplaces.

Live Load Variable weights imposed on building structure by snow, furniture, and people.

Lockset Door-knob-and-latch assembly, including provision for locking the door.

Louver Slanting overlapping slats of wood or metal fitted into framework for ventilating use.

Luminous Ceiling Translucent ceiling material with fluorescent back-lighting.

Masonry Construction using brick, stone, concrete, or clay tiles, usually done by a mason.

Meter Electric, gas, or water measurement of service of a dwelling.

Miter Cutting on an angle, usually with moldings or trim.

Mortar Portland cement, hydrated lime, sand, and water mixed to embed masonry units.

Mullions Vertical bars or moldings between window units.

Nonmetallic Cable Group of two or more electrical conductors within a sheath of nonmetallic material, different from metallic-armored cable.

O.C., On Center Indicates spacing of framing members.

Outlets Electrical-system access points for lighting or appliance use.

Overhang Outward projecting eave-soffit area of a roof.

Packaged Chimney Fabricated, metal chimney in sections, with necessary installation accessories.

Panels Either individual sheets or a flat assembly of parts, such as a stud wall.

Particleboard Sheet material made from compressed wood chips, flakes, or small particles.

Pitch Angle or slope of a roof, slope of a pipe, and a common name for roofing bitumen.

Plates Top one or two framing members running horizontally atop a wall; also the single plate on which wall studs rest.

Plenum Enlarged portion of the duct system adjacent to furnace or conditioner.

Plot Plan Drawings of the homesite or lot showing placement of the house and other details.

Plumb Straight up and down, measured with a level or plumb bob.

Polyethylene Plastic ingredient for clear waterproof film material used as a protective covering or as a vapor barrier.

Prefabricated Component or subassembly prepared in the factory or shop, saves on-job erection time.

Prehung Doors Door-and-frame assemblies, hinges mounted, holes for locksets drilled, thresholds in place, and sometimes prefinished.

Rafters Framing members that support the roof.

Rake Sloping edge of a roof often trimmed with a rake board matching the eave fascia.

Receptacles Electrical term for box-mounted devices that receive plug-in cords.

Registers Heat distribution devices that fit on the room end of heat ducts.

Resawn Unmilled or rough surface showing saw marks on rustic shingles or sidings.

Ridgeboard Board between upper ends of rafter pairs to frame the ridge of the roof.

Roof Decking Sheets or boards for sheathing over rafters.

Roofing Felt Asphalt building paper applied to roof sheathing before shingles.

Rough-in Measurements In plumbing, to show fixture pipe locations; also used for other built-in accessories or appliances.

Rough Opening Opening in building frame for windows, doors, stairs, chimney.

Roughing-in Installation of a system's parts inside construction framing before wall finish.

Scaffold Temporary equipment or framework to provide an elevated working platform.

Screed Horizontal wood or steel member staked to ground to provide guide for top surfacing of concrete slab.

Service Entrance Electric wiring run from utility company pole or transformer to house; also called Service Drop.

Setback Minimum distance a municipality requires from front of home to front property line.

Sheathing Boards or sheets applied over framing members to brace them and to serve as base for applying roofing and siding materials.

Shoe Thin molding used at bottom of baseboard and often called Base Shoe.

Shop Drawings More detailed and larger-scale drawings that supplement the construction set.

Sill Plate Bottom horizontal member of an exterior wall frame which rests atop foundation, sometimes called Mudsill.

Slab Horizontal concrete such as floor or drive.

Soffit Underside of a roof overhang.

Soil Pipe Pipe carrying toilet and other fixture wastes to the sanitary sewer.

Soil Stack Vertical soil pipe that acts as a drain at bottom and vent at top.

Solderless Connector Hollow screw-on device to fasten electric conductors together, sometimes called Wire Nut.

Sole Plate Bottom member of interior wall frame.

Specifications Written elaboration in specific detail about construction materials and methods, supplements working drawing.

Starter Strip Strip of preparatory material in first courses of roofing or siding.

Stringers Side supports for stair treads.

Studs Vertical framing members of a wall.

Subfloor Rough floor deck of sheets or boards over which finish floor is laid.

Tape-Cement Joints Joint treatment for drywall panels.

Three-Way Switch For controlling light from two locations.

Threshold Sill or bottom part of a doorway.

Tiles Building material units of clay, cement, plastic, or vinyl applied to flat surfaces, wall structures, or waste-drain lines.

Toe-nail Nailing through one member at an angle and into another.

Tongue-and-Groove Joint treatment where one strip or sheet edge fits into edge of another; with wood boards, called Dressed-and-Matched.

Top Plate Upper horizontal member of a wall frame.

Trap Fitting for plumbing fixture drainpipes to prevent sewer gas from entering the home by means of a water seal.

Treads Boards that form steps in a stairway.

Trim Wood, plastic, or metal finishing strips.

Troweling Final smoothing finish for concrete.

True A term for plumb.

Underlayment Smooth-surface sheet material applied under floor coverings.

Valley Framing Forming the angle between two inclined roof planes.

Vapor Barrier Sheet of vaporproof material to stop the passage of moisture vapor.

Vent Pipe Plumbing pipes that allow sewer gas to escape through the roof and also help to prevent siphonage and back pressure.

Ventilators Air exhaust or intake passage devices.

Wall Assemblies Components for wood-frame wall construction.

Wall Jack A pole-and-pulley assembly using rope to raise a section of wall too heavy to be lifted in place by hand.

Wall Plate Upper and lower horizontal members of a wall frame; also the finishing cover for an electrical switch or outlet.

Warm-Air System Heating system with either gravity flow or forced flow by electric blowers.

Waste Pipe Plumbing pipe to handle waste water.

Watt Electrical unit which measures the current drain of electrical equipment; one ampere of drain at one volt equals one watt.

Weather Stripping Narrow strips of wood, plastic, fiber to stop air flow at door and window.

Working Drawings Final construction drawings to show how buildings should be erected.

Index